CLOSTRIDIUM BOTULINUM

The Practical Food Microbiology Series has been devised to give practical and accurate information about specific organisms of concern to public health. The titles in the series are:

E. coli
Listeria
Clostridium botulinum
Salmonella

CLOSTRIDIUM BOTULINUM

A practical approach to the organism and its control in foods

Chris Bell
Consultant Food Microbiologist
UK

Alec Kyriakides
Company Microbiologist
Sainsbury's Supermarkets Ltd
London, UK

Blackwell
Science

© 2000 C. Bell and A. Kyriakides

Blackwell Science Ltd
Editorial Offices:
Osney Mead, Oxford OX2 0EL
25 John Street, London WC1N 2BL
23 Ainslie Place, Edinburgh EH3 6AJ
350 Main Street, Malden
  MA 02148 5018, USA
54 University Street, Carlton
  Victoria 3053, Australia
10, rue Casimir Delavigne
  75006 Paris, France

Other Editorial Offices:

Blackwell Wissenschafts-Verlag GmbH
Kurfürstendamm 57
10707 Berlin, Germany

Blackwell Science KK
MG Kodenmacho Building
7-10 Kodenmacho Nihombashi
Chuo-ku, Tokyo 104, Japan

First published 2000

Set in 10.5/12.5 pt Garamond Book
by DP Photosetting, Aylesbury, Bucks
Printed and bound in Great Britain by
MPG Books Ltd, Bodmin, Cornwall

The Blackwell Science logo is a trade mark of
Blackwell Science Ltd, registered at the United
Kingdom Trade Marks Registry

DISTRIBUTORS

Marston Book Services Ltd
PO Box 269
Abingdon
Oxon OX14 4YN
(*Orders:* Tel: 01235 465500
         Fax: 01235 465555)

USA
  Blackwell Science, Inc.
  Commerce Place
  350 Main Street
  Malden, MA 02148 5018
  (*Orders:* Tel: 800 759 6102
              781 388 8250
         Fax: 781 388 8255)

Canada
  Login Brothers Book Company
  324 Saulteaux Crescent
  Winnipeg, Manitoba R3J 3T2
  (*Orders:* Tel: 204 837 2987
         Fax: 204 837 3116)

Australia
  Blackwell Science Pty Ltd
  54 University Street
  Carlton, Victoria 3053
  (*Orders:* Tel: 03 9347 0300
         Fax: 03 9347 5001)

A catalogue record for this title is available
from the British Library

ISBN 0-632-05521-9

Library of Congress
Cataloging-in-Publication Data
Bell, Chris.
    Clostridium botulinum: a practical
approach to the organism and its control in
foods/Chris Bell, Alec Kyriakides.
        p. cm.—(Practical food microbiology
series)
    Includes bibliographical references and
index.
    ISBN 0-632-05521-9 (pbk.)
    1. Botulism.   2. Food—Microbiology.
3. Clostridium botulinum.   I. Kyriakides,
Alec.   II. Title.   III. Series.

QR201.B7 B37 1999
616.9'315—dc21   99-049855

For further information on
Blackwell Science, visit our website:
www.blackwell-science.com

CONTENTS

FOREWORD

Although botulism is rare compared to many other microbial diseases, the disease has long fascinated physicians and scientists because of the unusual symptoms and relatively high fatality rate of affected humans and animals. Foodborne outbreaks of botulism invariably generate considerable publicity because the organism responsible (*Clostridium botulinum*) has the ability to synthesise a neurotoxin of extraordinary potency.

The meticulous studies of Kerner from 1815 to 1828 and van Ermengem from 1895 to 1897 provided an invaluable insight into the disease, the organism and the properties of its toxins. Whole conferences and books have subsequently been devoted to these topics. With so much information available there are inherent problems in reducing this to manageable dimensions. Chris Bell and Alec Kyriakides, both very experienced microbiologists, have undertaken the difficult task of selecting and condensing all the available information into what really matters and what, like good food, can easily be digested.

The book is divided into seven chapters complemented by over 80 tables, numerous figures and a detailed reference list and glossary of terms. The uniqueness of this volume is that it provides a truly user-friendly, practical approach to the organism and its control in food.

Having been personally involved in the laboratory investigation of two major outbreaks of botulism in this country I have enjoyed reading this book, learning so much new and understanding and appreciating what I should have known already. Botulism, like all foodborne diseases, can and should be prevented by creating, applying and verifying good manufacturing practice – a strong theme throughout the book.

In summary, this book is very comprehensive and admirably concise and the food industry should take heed of the practical advice and solutions it contains.

Professor Richard J. Gilbert OBE
Formerly Director - Food Hygiene Laboratory
PHLS Central Public Health Laboratory
London

1

BACKGROUND

INTRODUCTION

The toxins produced by *Clostridium botulinum* are among some of the most potent, naturally occurring toxic substances known. If spores of *Clostridium botulinum* are present in food and conditions are not inhibitory to germination and growth, the organism can proliferate and produce toxin. Such occurrences have been the cause of outbreaks of botulism in the human population. Botulism is a severe illness in which the nervous system is affected and mortality rates can be high due to respiratory failure.

The organism is widespread in soil and aquatic sediments and also in the gastrointestinal tracts of animals, fish and birds. There is, therefore, a significant potential for the organism to be present in or on raw foods. Because the potential consequences of the growth of *Clostridium botulinum* in foods are so serious, the food industry must be aware of the hazard and the measures necessary for its control.

The illness caused, botulism ('sausage poisoning'), is so called because it had, for many years, been associated with the consumption of sausages. According to Mitchell (1900), cases of botulism were most frequent in parts of Germany where raw sausage and raw ham were most widely consumed. He noted several outbreaks in which large numbers of cases were involved. For example, in 1879, 241 people in Chemnitz were poisoned by Mettwurst; in 1886, 160 individuals, again in Chemnitz, were poisoned by Mettwurst; after 1886, 11 cases occurred in Dresden, more than 50 cases in Gerbstadt and 30 cases in Gera.

Although the symptoms of botulism were well known, mortality associated with the illness was known to be high, and early studies by Kerner identified blood and liver sausages as the main causes of botulism documented between 1815 and 1828 (234 cases and 110 deaths) (Hauschild, 1989). However, the causative agent of the illness was not discovered until an outbreak which occurred in 1895 was investigated. In December of that year, in the Belgian village of Ellezelles, several individuals were poisoned through eating a particular pale, soft and rancid-smelling portion of a raw ham; other ham and the rest of the animal had been consumed without causing ill effects (Topley and Wilson, 1929). Four people died and from the liver of one of the victims, van Ermengem isolated an anaerobic bacillus, cultivations of which produced the symptoms of 'sausage poisoning' when inoculated into animals. Subsequently, a 'virulent toxine (*sic*)' was isolated from the bacterial cultures which was deemed to be the direct cause of the illness (Mitchell, 1900). The organism isolated was named *Bacillus botulinus*.

An organism reported to have been isolated from swine faeces by Kempner in 1889 was identical in morphological and pathogenic properties to *B. botulinus*. In addition, the toxin from this isolate produced the symptoms of botulism. This was regarded as an indication of the origin of van Ermengem's organism (Mitchell, 1900). It was also Kempner who reported in 1897 that goats could be made immune to the toxin of *B. botulinus* by injecting them with gradually increasing doses of toxin. It was also found that guinea pigs treated with the blood serum of the immune goats could withstand doses of toxin 100 000 times the normally fatal dose.

This early work was rapidly complemented over the following 2–3 decades with research investigating the characteristics of the organism and its toxins, the symptoms and pathology of human botulism, the distribution of the organism in the environment, and routes of food contamination. Reasons for the involvement of specific food types in outbreaks became known and potential treatments of the illness were developed using antitoxin to toxin types A and B. During this period, the organism's role in botulism-like illnesses in poultry (limberneck), horses (forage poisoning), cattle (bulbar paralysis/forage poisoning) and other animals was also being investigated (Topley and Wilson, 1929).

By the time of the Loch Maree incident in Scotland in August and September, 1922, in which eight people died after eating sandwiches made with wild duck paste, Leighton, one of the investigators who wrote a detailed account of the incident (Leighton, 1923) also reviewed the considerable body of knowledge that had accumulated by then. Much of this work had been done in California, USA, where many outbreaks had occurred due to poorly home-preserved foods. Meyer provided Leighton with information about cases and deaths from botulism in the USA up to December 1922 which showed a mortality rate of 62.8% (110 outbreaks/ 396 cases/249 deaths). Foods involved included asparagus in jars, olives in jars, canned spinach, cottage cheese (Nevin, 1921), olive relish, canned beets and tomato-onion-chilli sauce, indicating a predominance of vegetable products implicated in USA outbreaks of botulism.

Today, *Clostridium botulinum* is universally recognised as an important potential hazard in a wide range of food types but the requirement for specific control measures in some food processing areas is sometimes not well understood or inadequately applied. The consequences of this situation are the continued reports, albeit at a lower incidence, of outbreaks of this most serious of foodborne illnesses, botulism.

TAXONOMY OF *CLOSTRIDIUM BOTULINUM*

The term 'botulism' derives from the Latin 'botulus' which means 'sausage', referring to the common term for the illness, 'sausage poisoning'. When the causative organism was first described by van Ermengem in 1896 (Topley and Wilson, 1929), the organism was named *Bacillus botulinus* (*Bacillus* deriving from the Latin 'bacillum', meaning a small rod).

The name *Clostridium* was first introduced by Prazmowski in 1880 to describe the spindle shapes of organisms observed at that time (due to the endospore being wider than the vegetative cell). *Clostridium* derives from the Greek 'closter', a spindle (Cato *et al.*, 1986), and this name was used subsequently by different workers for various species of the spore-bearing family Bacillaceae (Fairbrother, 1938). It was only in 1917 that 'The American Committee' designated all spore-bearing anaerobes to the genus *Clostridium*, the description given as: 'anaerobes; often parasitic; rods frequently enlarged at sporulation, producing clostridium or plectridium forms' (Fairbrother, 1938). Today, the name *Clostridium botulinum* (*C. botulinum*) (although a misnomer now because the 'sausage poisoning' connection is historical) is internationally understood and accepted to encompass organisms that produce a botulinum neurotoxin.

The genus *Clostridium* is a member of the family Bacillaceae and members of the genus are distinguished from other Gram-positive endospore-forming genera of the family by requiring anaerobic conditions for growth and being, usually, catalase negative (Cato *et al.*, 1986). The species *C. botulinum* is motile with peritrichous flagellae and has cells in the range 0.3–0.7 × 3.4–7.5 µm in size (International Commission on Microbiological Specifications for Foods, 1996). Based on the serological specificity of the toxins produced, seven types of *C. botulinum* are currently recognised (A–G) and, using these serological and cultural characteristics, the species has been separated into phenotypic subgroups I–IV (Table 1.1). Results from the application of modern molecular techniques have demonstrated that the four phenotype groups also correspond genetically.

Table 1.1 Characteristics of the subgroups of *C. botulinum*, adapted from Cato *et al.* (1986) and Hauschild (1989)

Properties	Group			
	I	II	III	IV
Neurotoxin type	A, B, F	B, E, F	C (alpha and beta), D	G
Proteolytic*	yes	no	no or only mildly so	weakly
Saccharolytic	no	yes	no	no
Lipolytic	yes	yes	yes	no
Psychrotrophic	no	yes	no	no
Associated with botulism in humans	yes	yes	rare	rare or doubtful

*Proteolysis = digestion of casein, serum proteins, meat or coagulated egg.

Of most particular concern with respect to food safety are the proteolytic *C. botulinum* types A and B in canned meats and vegetables and non-proteolytic types B and E in fish products and, increasingly, cooked-chilled products.

Until recently, despite differences in other characteristics, all organisms demonstrated to produce botulinum neurotoxin have been included in the genus *C. botulinum*. However, other species of *Clostridium*, notably some strains of *C. butyricum* and *C. baratii*, have been found to produce botulinum toxins type E and type F respectively, and *C. botulinum* type G (Group IV) has been re-named *C. argentinense* (International Commission on Microbiological Specifications for Foods, 1996).

Clearly, the identification and grouping of botulinum toxin-producing *Clostridium* spp. will be an ongoing area of study. Of particular importance to the food industry is information concerning not only the identity of the organism but also its source and the circumstances involved in the cause of any foodborne outbreak of botulism.

BOTULISM: THE ILLNESS

Human foodborne botulism is usually an intoxication where illness follows the ingestion of pre-formed toxin in food. The toxins produced by *C. botulinum* and affecting humans are neurotoxins which attack the nervous system of the affected individual. They block the release of acetylcholine, a neurotransmitter at the peripheral nerve ends. Transmission of nerve impulses at the neuromuscular junctions is prevented and no muscle stimulation occurs, resulting in flaccid paralysis. The mechanism by which this occurs has in recent years been found to be highly specific for each toxin type. The nature and properties of botulinum neurotoxins and their mode of action are subjects of study in their own right and further information may be found in Smith (1977), Sugiyama (1980), Shone (1987) and Hauschild (1989).

Various estimates have been made of the amount of botulinal toxin required to cause toxic effects in humans and the lethality of some of these toxins. For example, the toxic effect of type E may be increased by trypsin, a protein-splitting enzyme secreted initially as trypsinogen by the pancreas and found as trypsin in digestive juices. This 'toxicity enhancement' possibly explains the high mortality rate observed in many type E botulism outbreaks despite its relatively low lethality in mice when compared to types A and B (International Commission on Microbiological Specifications for Foods, 1996). Toxic doses for humans of all botulinum toxin types, however, are estimated to be very low, i.e. at levels of < 1 µg for toxin types A and B and approximately 10 µg for toxin types E and F. As a consequence, assays for the detection of toxin in foods have to be reliably sensitive to the presence of extremely low levels of the toxin. It is also the reason why the mouse assay, which provides a measure of the biological activity of toxins as well as being sensitive to levels of around 0.01–0.02 ng of toxin types A or B per ml and 0.1 ng/ml of toxin types E or F, is still widely used (Sugiyama, 1980; Shone, 1987).

Three categories of botulism are recognised in humans; in addition to foodborne botulism, there are also wound botulism and infant botulism. Table 1.2 indicates the general symptoms associated with each of these categories.

Table 1.2 Cause and symptoms associated with different categories of botulism

Category	Botulinum type	Cause and symptoms	Reference
Wound	A and proteolytic type B	Normally rare but may be seen in wartime situations. Organism in the environment, e.g. soil, gains access to and colonises deep wounds where there are anaerobic conditions caused by tissue destruction and damaged blood supply. Toxin is produced which gets into the circulatory system and neurological symptoms (not gastro-intestinal) similar to those in foodborne botulism occur.	Cato *et al.* (1986) Salyers and Whitt (1994)
Infant	A B E (*C. butyricum*) F (*C. baratii*)	Spores of *C. botulinum* ingested, e.g. in honey, by the infant (usually <6 months of age) germinate and multiply, producing toxin in the gut. The toxin is distributed in the blood stream reacting at nerve/muscle junctions and may ultimately cause flaccid paralysis (floppy baby syndrome) and death.	Arnon (1980) Hauschild (1989)
Foodborne	A B E C, D and F – rare	*C. botulinum* grows in food and produces toxin which is ingested with the food. Incubation period varies from a few hours up to 8 days but usually 18–36 hours. Symptoms may include nausea, vomiting, diarrhoea, muscle weakness, fatigue and dizziness followed by disturbances to vision, difficulty in swallowing and talking, paralysis, respiratory difficulty or failure causing death.	Smith (1977)

Botulism may be misdiagnosed because symptoms resemble other ill-nesses. Early misdiagnosis can lead to a delay in recognising an outbreak, such as that which occurred in 1985 in British Columbia, Canada, due to consumption of contaminated and toxic chopped garlic in soybean oil. Prior to the identification of three cases of botulism which led to the ultimate discovery of 33 further cases, 28 of these had been diagnosed with other illnesses including stroke, myasthenia gravis, viral syndrome and psychiatric illness, amongst others (St Louis, 1988). A variety of other misdiagnoses have also been recorded such as food poisoning caused by other bacteria, including *Salmonella* spp. and *Clostridium perfringens*, tick paralysis, cerebro-vascular accidents, non-microbial poisoning including mushroom, belladonna, atropine, carbon monoxide and some other conditions (Smith, 1977). Consequently, the incidence of botulism around the world is likely to be under-reported.

In foodborne botulism, incubation times, symptoms and severity of illness including likelihood of death can vary considerably from individual to individual depending on toxin type, amount ingested and whether or not vomiting occurs which could eject some of the toxic food. Toxin types B and E tend to produce symptoms more quickly than type A and higher mortality rates have tended to be associated with high toxin doses which produce a rapid onset of symptoms (Smith, 1977). The mortality rate associated with botulism can be high with toxin types A and E but tends to be lower in outbreaks caused by non-proteolytic type B, where the illness seems to develop more slowly. Treatment of botulism involves prompt administration of antitoxin, often polyvalent antitoxin because the spe-cific causative type may not be known, and good medical care providing artificial respiration facilities if required. A significant reduction in mor-tality has been observed since such treatments have been applied, for example in Russia where botulism of types A and B were most common. Between 1818 and 1913 there were 609 cases with 283 deaths (46.5% fatality) whereas between 1920 and 1939, when antitoxins to these two types were generally available, there were 674 cases with 176 deaths (26.1% fatality) (Smith, 1977).

Botulism also occurs in animals and birds and Table 1.3 summarises botulinum toxin types and some associated susceptible groups.

The botulinum toxin types important in adult human foodborne botulism are A, B, E and F, although type F is believed to be rare; types C and D have also been very rarely reported in human foodborne cases. Botulism occurs throughout the world although some countries, such as Australia, India, Sweden, Switzerland and the UK, appear to have a very low incidence

Table 1.3 Some groups susceptible to different types of botulism, adapted from Smith (1977), Hauschild (1989) and International Commission on Microbiological Specifications for Foods (1996)

Botulinum type	Affected groups
A	Man, infants, waterfowl
B	Man, infants, horses
C alpha	Waterfowl, chickens
C beta	Cattle, sheep, horses
D	Cattle, horses
E	Man, fish, infants
F	Man, infants
G	Unclear

(Smith, 1977; Hauschild, 1989). Different types of *C. botulinum* have been found to predominate and cause outbreaks largely according to climate and diet (Table 1.4).

Table 1.4 Predominant types of botulism in different parts of the world

Botulism type	Some regions in which type predominates	Main associated foods
A	Argentina, China, USA	Vegetables
B	Spain, Poland, France, Germany, Czech Republic, Belgium	Meats
E	Japan, Canada, Alaska (USA), Denmark, Iran	Fish

SOURCES OF *CLOSTRIDIUM BOTULINUM*

Since the identification of *C. botulinum* as the causative organism of botulism, many investigations have been carried out into its distribution in the environment, the routes of food contamination and the factors contributing to the development of toxin in foods leading to outbreaks of botulism.

By 1922, Meyer and Dubovsky had examined soil samples from many parts of the USA, in addition to vegetables and fodders, and found that up to 30% of samples were positive for the organism (Topley and Wilson, 1929). Although toxin type A was found to predominate, types A and B were both recovered from all soil types examined. They also examined soil samples from some European countries, including England, Belgium, the Netherlands, Denmark and Switzerland, and found type B but not type A in a variable proportion of the samples from each country. However, *C. botulinum* types A and B were found in soil samples from Canada and China (Topley and Wilson, 1929). Since this early work, many more surveys have been carried out to assess the environmental distribution of the different toxin types in both land soils and aquatic bottom deposits. Table 1.5 summarises results from some of this work. It is clear that the organism is a widespread natural inhabitant of both land and aquatic soils/mud and that, although distribution is uneven within any selected area, different toxin types can be seen to predominate in particular regions.

Environmental survey work has extended to surveys of the extent of carriage of *C. botulinum* in animals, birds, fish and other marine creatures. Early work found that the organism was occasionally present in the faeces of cattle, pigs and horses, but rarely in healthy human faeces (Topley and Wilson, 1929). Vegetables including string beans (pods and stalks), corn husks, leaves and stalks, beets (roots and tops) and tomato plants and roots were examined for the presence of *C. botulinum* by Meyer and Dubovsky in the USA. Positive findings were recorded for all vegetable types and the incidence was found to be in the range 7.5–31.8% (Topley and Wilson, 1929). In the same work, different animal fodders were also found positive for the organism, the incidence ranging from 15 to 20%.

The now extensive studies that have been carried out to assess the incidence of *C. botulinum* in fish and aquatic invertebrates have reflected the findings in aquatic sediments. The toxin types identified in fish and aquatic invertebrates, particularly bottom feeders and scavengers such as crabs and prawns, bear a close association with those found in the environment (Table 1.6). Huss (1980) examined soils and sediments from a wide variety of Danish localities; his results suggested that *C. botulinum* type E is a true aquatic organism and is not derived from land soils. As its presence did not appear to correlate with the location of industrial pollution or the presence of large quantities of rotting vegetation it was considered unlikely that the organism multiplied in the bottom deposits but most probably in dead fish and other aquatic fauna. The presence of a rich aquatic life was considered important in contributing to the incidence of *C. botulinum* type E in the aquatic environment.

As soil and sediments have been found to be significant natural sources of *Clostridium botulinum*, it should be expected that from time to time all

Table 1.5 Examples of some survey results indicating the environmental distribution of different *C. botulinum* toxin types in land soils and aquatic bottom deposits

Location	Sample type	% sites or samples positive	Toxin types detected	Types as % of positive sites/ samples	Reference
England and Wales	Soils	13.1	B and untyped	B – 55.5	Meyer and Dubovsky (1922a)
Switzerland	Soils and vegetables	35.2	B and untyped	B – 72.7	Cann et al. (1965)
North Sea	Bottom deposits	0	E	0	Cann et al. (1965)
Coast of Scandinavia	Bottom deposits	79.0	E	100	Cann et al. (1968)
British coastal waters	Bottom deposits	3.5	B	100	Cann et al. (1968)
Denmark	Marine sediment	92.0	E	>90	Huss (1980)
	Freshwater sediment	86.0	E	>90	
	Cultivated farmland	41.0	B	>90	
Great Britain	Mud from aquatic environments in England, Wales and Scotland	36.9	B C D E	79.7 9.6 3.1 7.6	Smith et al. (1978)
Ireland	Mud from aquatic environments	18.0	B	100	
Great Britain	Soils across England, Wales and Scotland	10.0	B	100	Smith and Young (1980)

Table 1.5 Continued

Location	Sample type	% sites or samples positive	Toxin types detected	Types as % of positive sites/samples	Reference
Eastern USA	Soil	19	A	12	Hauschild (1989)
			B	64	
			C/D	12	
			E	12	
Newfoundland,	Aquatic sediments	9.7	A	0	
Gulf of St Lawrence,		18.2	B	0	
Gulf of Maine		0.6	C/D	0	
			E	100	
Western USA	Soil	28.0	A	62	
			B	16	
			C/D	14	
			E	8	
Alaska	Aquatic sediments	48.5	A	0	
			B	0	
			C/D	0	
			E	100	
Argentina	Soil	33.8	A	67	
			B	20	
			C/D	0	
			E	0	
			F/G	5/1	

Table 1.5 Continued

Location	Sample type	% sites/samples positive	Toxin types detected	Types as % of positive sites/samples	Reference
Poland, Baltic coast	Soil	31.5	A	3	Hauschild (1989)
			B	0	
			C/D	0	
			E	97	
Former USSR (European)	Soil	9.4	A	8	
			B	17	
			C/D	2	
			E	73	
Former USSR (Asian)	Soil	12.6	A	9	
			B	44	
			C/D	2	
			E	44	

Table 1.6 Examples of the relationship between *C. botulinum* types found in fish and their environment

Location	Sample type	% of samples positive	Toxin types as % of positive samples		References
Gulf of Maine	Aquatic sediments	0.6	A	0	Hauschild (1989)
			B	0	
			C/D	0	
			E	100	
	Fish	4.5	A	0	
			B	0	
			C/D	0	
			E	100	
East Coast USA, New York to Florida	Aquatic sediments	3.6	A	8	
			B	17	
			C/D	42	
			E	33	
	Fish and invertebrates	0.8	A	0	
			B	33	
			C/D	33	
			E	33	
Alaska	Aquatic sediments	48.5	A	0	
			B	0	
			C/D	0	
			E	100	

Table 1.6 Continued

Location	Sample type	% of samples positive	Toxin types as % of positive samples		References
Alaska	Salmon	4.9	A	0	Hauschild (1989)
			B	0	
			C/D	0	
			E	100	
Scotland	Aquatic sediments	26.0	B	90	Smith *et al.* (1978)
			C/D	10	
	Trout	1.0	B	100	Dodds (1993a)
			C/D	0	
Off the coast of Scandinavia	Bottom deposits	79.0	E	100	Cann *et al.* (1965)
Sweden – The Sound	Fish	100	E	100	Dodds (1993a)
West Indonesia coast	Aquatic sediments	1.9	A	18	Hauschild (1989)
			B	36	
			C/D	46	
			E	0	
			F	0	
	Fish	2.7	A	20	
			B	16	
			C/D	54	
			E	0	
			F	10	

Table 1.6 Continued

Location	Sample type	% of samples positive	Toxin types as % of positive samples		References
Thailand, Gulf of Siam	Aquatic sediments	1.6	A	0	Hauschild (1989)
			B	0	
			C/D	83	
			E	17	
	Fish	0.7	A	0	
			B	0	
			C/D	67	
			E	33	
Iran, Caspian Sea coast	Aquatic sediments	17.0	A	0	
			B	8	
			C/D	0	
			E	92	
	Fish	17.7	A	0	
			B	4	
			C/D	0	
			E	96	
Finnish trout farms	Sediments	68	A	0	Hielm et al. (1998)
			B	0	
			E	100	
			F	0	
	Fish intestines	15	A	0	
	Fish skins	5	B	0	
			E	100	
			F	0	

raw food types will be contaminated with the organism. Dodds (1993b) summarised the results from a variety of published surveys carried out to determine the incidence and level of *C. botulinum* spores in different foods (Table 1.7). In addition, Lilly *et al.* (1996) found spores of *C. botulinum* in four packs from an examination of over 1100 commercially available, modified atmosphere, packaged prepared vegetables (one each of shredded cabbage (type A), chopped green pepper (type A), Italian salad mix (type A) and escarole salad mix (types A and B)).

Where levels of spores present in foods have been estimated, these have generally been found to be low in meats, < 10 spores per kg, and often < 1/kg, but highly variable in fish (< 1–2400/kg) and vegetables (< 1–2100/kg) (Dodds, 1993b). The contamination level in raw milk has been estimated to be < 1 spore/litre (Collins-Thompson and Wood, 1993).

Because of the apparent link between infant botulism and consumption of honey, a great deal of work has been carried out to determine the incidence of spores in honey. Levels reported range from < 1 to 35 spores/kg in honey not associated with illness, but in two incidents (Canada and the USA) where the honey examined was associated with illness, levels were estimated to be 8000 and 80 000/kg, respectively (Dodds, 1993b). *C. botulinum* types A and B are those most commonly found in honey in the USA, Europe, Canada, Japan and China but type C has also been found in honey in Japan, China, Hungary, Mexico and Argentina; type D has been recovered from honey in Argentina.

The widespread nature of the organism in soil and aquatic environments leads to widespread contamination of raw food materials which, in turn, accounts for the broad spectrum of foods that have been associated with outbreaks of botulism. In an analysis of outbreaks in which the food vehicle was identified, 90% or more of such outbreaks in Alaska, Canada, former Czechoslovakia, Hungary, Japan, Spain, the USA and former USSR were due to the consumption of poorly processed and stored home-preserved foods (Hauschild, 1989). The main food types involved often reflected the dietary habits of the nation in question, for example, cases in Alaska, Canada and Japan mainly involved fish and marine mammals (Table 1.8).

Outbreaks of botulism caused by commercially produced foods, although infrequent, are of particular importance because of the possibility of widespread distribution of the foods both nationally, within the producing country, and internationally. The possible difficulties of diagnosis of this serious illness could lead to significant mortality rates in those affected. Table 1.9 indicates a range of commercially produced food types,

Table 1.7 Incidence of *C. botulinum* spore types in foods

Food type	Source	% positive	*C. botulinum* type	Reference
Prepared fish				
Haddock fillets	North America, Atlantic	24	E	Dodds (1993b)
Dressed rockfish	North America, California	100	A, E	
Salmon	North America, Washington	8	E	
Salmon	Alaska	100	A	
Smoked herring	Sweden	13	E	
Smoked salmon	Denmark	2	B	
Salted carp	Caspian Sea	63	E	
Shrimp	Indonesia	2	B, C, F	
Crab	Indonesia	7	A, C, D	
Smoked eel	Baltic Sea	20	E	
Meat and poultry				
Raw pork	UK	0–14	A, B, C	Dodds (1993b)
Vacuum-packed bacon	UK	4–73	A, B	
Raw meats	North America	<1	C	
Raw meats	Europe	36	E	
Fruits and vegetables				
Fruits and vegetables	USA	13	A; B	Dodds (1993b)
Vegetables, herbs	Former USSR	43	A; B	
Vegetables	Italy	4	B	
Precut MAP vegetables	USA	0.36	A or A and B	Lilly *et al.* (1996)
Fresh mushrooms	Netherlands	estimated <0.08–0.16 organisms/100 g	B (found in casing soil of production beds)	Notermans *et al.* (1989)

Table 1.8 A summary analysis of some outbreaks of botulism (country, cases, foods and types), adapted from Hauschild (1993)

Country	Years	Total outbreaks/ Number of outbreaks in which type identified/Food identified	Total cases (deaths, %)	Main botulinum types of those identified (%)			Food types (% of those identified)				Home-made (%)	Commercially produced (%)
				A	B	E	Meats	Fish	Fruits and vegetables	Other		
USA	1971–1989	272/252/222	597 (11)	61	21	17	16	17	59	9	92	8
Canada	1971–1989	79/76/75	202 (14)	4	8	88	72	20	8	0	96	4
Hungary	1985–1989	31/31/28	57 (2)	0	100	0	89	0	4	7	100	0
France	1978–1989	175/171/123	304 (2)	0	97	2	89	3	6	2	88	12
Spain	1969–1988	63/36/48	198 (6)	0	92	3	38	2	60	0	90	10
Denmark (inc. Greenland)	1984–1989	11/11/10	16 (12)	0	0	100	100	0	0	0	100	0
Norway	1961–1990	19/19/19	42 (7)	0	47	47	16	84	0	0	100	0
Former USSR	1958–1964	95/45/83	328 (29)	33	38	29	17	67	16	0	97	3
China	1958–1983	986/733/958	4377 (13)	93	5	1	10	0	86	4	?	?
Japan	1951–1987	97/97/95	479 (23)	2	2	96	0	99	1	0	98	2
Argentina	1980–1989	16/13/14	36 (36)	77	8	0	29	21	36	14	79	21
UK	1955–1989	3/3/3	33 (9)	33	33	33	0	66	0	33	0	100
Australia	1942–1989	5/1/5	53 (17)	20			0	20	80	0	0	100
Brazil	1958–1990	4/4/3	19 (47)	100			25	25	25	25	50	25?
Kenya	up to 1979	2/1/2	17 (65)	50			0	0	0	50* / 50†	100	0

Note: information concerning outbreaks is not always complete so percentages given are those calculated from data available; therefore, figures will not always add up to 100%.
* Sour milk
† Termites
? Not known/not reported

Table 1.9 Some outbreaks of botulism involving commercially produced foods (including restaurants)

Year	Country	Food	Cases	Deaths	Botulism type	References
1922	Scotland	Potted duck paste	8	8	A	Leighton (1923)
1977	USA	Canned jalapeno peppers	59	0	B	Terranova et al. (1978)
1978	UK	Canned Alaskan salmon	4	2	E	Ball et al. (1979)
1978	USA	Potato salad	7	0	A	Seals et al. (1981)
1983	USA	Sauteed onions	28	1	A	MacDonald et al. (1985)
1985	Canada	Bottled chopped garlic in soybean oil produced in USA	36	0	B	St Louis et al. (1988)
1986	Taiwan	Heat preserved unsalted peanuts in jars	9	2	A	Chou et al. (1988)
1987	USA and Israel	Kapchunka (dry-salted, air-dried, uneviscerated whole white fish)	8	1	E	Telzak et al. (1990)
1989	UK	Hazelnut yoghurt (contaminated and toxic canned hazelnut conserve component)	27	1	B	Critchley et al. (1989) O'Mahony et al. (1990)
1993	Italy	Roasted eggplant in oil	7	0	B	D'Argenio et al. (1995)
1993	USA	Canned cheese sauce	8	1	A	Townes et al. (1996)
1994	USA	Dips containing baked potato	30	0	A	Angulo et al. (1998)
1996	Italy	Mascarpone cheese (acidified dairy cream)	8	1	A	Aureli et al. (1996)
1998	Italy	Vegetable soup in glass jars	1	0	A	Anon (1998a)

including some foods produced in restaurants, that have been implicated in outbreaks of botulism.

The range of food types and technologies of production represented in all these outbreaks is extensive. Home preserved foods, particularly vegetables in oil or brine, continue to be of concern (Anon, 1998b). As food processing and packaging technologies develop further and the shelf life of chilled foods in this still rapidly growing food sector is extended, there is an increasing need for food microbiologists and technologists to ensure that due diligence is maintained throughout the hazard analysis process for new products and technologies. This will help to ensure that any possible risk from *C. botulinum* is minimised.

If there is the potential for spores of *C. botulinum* to be present in a food where there is little or no competition from other organisms and where conditions in the food and storage conditions do not preclude spore germination, growth and toxin production, that food is potentially hazardous to the consumer. This is the set of circumstances which have combined to cause the majority of outbreaks, whether home or commercially produced, and explains why preserved, long-life foods such as hams, sausages and canned/bottled products are most often involved in outbreaks. The spores of the organism are present in the raw food and the process, often heating, kills most of the competitive organisms but is inadequate to kill the *C. botulinum* spores. In addition, the formulation of the product (which will often be stored for long periods because it is 'preserved') does not or may not preclude the growth of the organism. Despite the apparent rarity of foodborne botulism from commercially produced products, such foods and the production processes involved demand the particular attention of food microbiologists and technologists to ensure that adequate control procedures are in place to prevent outbreaks of botulism occurring.

The spores of *C. botulinum* must be expected to be present in, or on, all raw foods derived from agricultural or horticultural land and aquatic environments and the organism is consumed regularly by everyone without causing harm. It is in circumstances of food processing failures and/or abuse allowing growth of *C. botulinum* and toxin production that most often lead to outbreaks of botulism. Because of the large scale nature of the primary production, manufacture, and national and international distribution of both raw materials and processed foods, consistently applied, reliable measures to control growth of *C. botulinum* are essential. It is the control of *C. botulinum* in such commercial scale food processes that is addressed in this book.

2

OUTBREAKS: CAUSES AND LESSONS TO BE LEARNT

INTRODUCTION

Outbreaks of botulism in the UK are very rare. However, a large number of outbreaks occur every year throughout the world. Table 1.9 summarises some of the recorded outbreaks that have occurred over the years and demonstrates the world-wide nature of the hazard of *C. botulinum*.

Outbreaks have implicated almost all sectors of the food industry and, although the majority have been associated with processed meat, fish and vegetable products, some outbreaks have also involved dairy products. Outbreaks that receive widest attention are those associated with processed products from the food manufacturing industry, but many incidents and outbreaks of botulism are associated with home-made or traditional products such as naturally fermented meats and fish made in small communities throughout the world (Table 1.8).

Outbreaks of botulism are extremely serious and often involve fatalities and it is imperative that the food industry employs all practical means to prevent foods from presenting a significant risk in relation to this most serious of hazards. Many of the outbreaks that have occurred in recent years provide significant opportunities to learn from mistakes made by others. The next section reviews some of these outbreaks, considers what may have gone wrong and identifies controls that could have been implemented to prevent them. In this way it is hoped that future processes will be devised and operated using a knowledge of factors that, if implemented, should avoid a recurrence of incidents of botulism from foods.

As many of the outbreaks that are reported are primarily investigated using a clinical approach, many of the fine details of the product and process are not recorded. Therefore, in these cases, the authors have used a speculative approach to expand on possible causes and controls.

MASCARPONE: ITALY

Dairy products are rarely implicated in foodborne botulism, but one such product was implicated in an outbreak in south-western and northern Italy in 1996 (Simini, 1996). A total of eight people (six male and two female) between the ages of 6 and 24 years suffered symptoms of botulism 12–24 hours after the consumption of the implicated product (Table 2.1). Symptoms included nausea, cephalagia (aching of the head) and vomiting followed by neurological symptoms, with all patients being hospitalised and supported using artificial ventilation. Although all patients were treated with trivalent antiserum (Aureli *et al.*, 1996), one patient, a 15-year-old boy, died. The botulism was caused by *C. botulinum* type A; the toxin was detected in two out of four serum samples and two out of seven rectal swab samples from affected patients. *C. botulinum* type A was also isolated from the faeces of all patients.

Table 2.1 Outbreak overview: mascarpone

Product type:	Acidified cream cheese
Year:	1996
Country:	Italy
Levels:	$> 10^5$/g
Cases:	8 (1 death)
Type:	*C. botulinum* type A (proteolytic)

Possible reasons
(i) Heat process insufficient to destroy spores of *C. botulinum*
(ii) Product characteristics incapable of inhibiting spore germination and growth of *C. botulinum* under conditions of elevated temperature
(iii) Temperature abuse during the manufacturing process, in distribution, at retail level or by the consumer (or a combination of these)

Control options*
(i) Heat processing of cream to destroy spores of proteolytic *C. botulinum*, e.g. UHT process
(ii) Effective cooling after processing ($< 5°$C within four hours)
(iii) Clear instruction to maintain product in refrigerated storage, i.e. 'Keep refrigerated'
(iv) Reduced shelf life to limit time over which temperature abuse could have occurred

*Suggested controls are for guidance only and may not be appropriate for individual circumstances. It is recommended that proper hazard analysis is carried out for every process and product to identify where controls must be implemented to minimise the hazard from *C. botulinum*.

The product implicated in the outbreak was mascarpone, an acidified cream cheese, which was consumed by all patients either alone or as an ingredient in home-made tiramisù. The mascarpone was reported by a number of affected individuals to have had an offensive odour. Samples of the implicated brands of mascarpone, purchased from local stores, were found to contain *C. botulinum* type A at levels in excess of 10^5/g together with type A botulinum toxin at a level of 1866 LD_{50}/g. Samples of tiramisù from two of the patients' homes and one unopened pack of mascarpone from the local shop used by the patients contained type A toxin at 125 LD_{50}/g and 10^5 *C. botulinum* spores/g (tiramisù) and 2495 LD_{50}/g and $> 10^5$ spores/g (mascarpone). Lower levels of spores of *C. botulinum* type A (4 spores/g) were detected in samples of mascarpone taken from different production codes collected from local stores and from the implicated manufacturing site.

The outbreak was limited to eight people due to recall of the implicated brands of mascarpone, all manufactured by a single producer, together with cessation of production. Although the mascarpone was exported to a number of countries, including Austria, Belgium, Spain and the USA, no cases of botulism were recorded outside of Italy. However, product recall notices were issued in export countries to prevent further incidents (Food and Drug Administration, 1996).

Mascarpone, although often referred to as a cheese, is not a traditional fermented milk product. It is manufactured by a very simple process involving the acidification of pasteurised cream/milk. Raw milk and cream are mixed to achieve the desired fat content and then pasteurised at high temperatures, ranging from 85°C for 10-30 minutes to ultra-high temperature processes of 110-130°C for several seconds. The cream/milk mixture is acidified by the addition of citric acid at elevated temperatures. The combination of a high temperature and the acidity serve to denature the milk proteins and form a curd. Some whey is removed and then the product is usually homogenised to form a smooth, creamy paste. This is usually hot filled into bulk packs of several kilogram quantities or into smaller pre-pack containers for direct sale to consumers.

Mascarpone is a fairly bland product and is often used as an ingredient in desserts. Although acidified, the retail product has a fairly high pH of 5.7-6.5, a moisture content of approximately 50% and, consequently, quite a high water activity of approximately 0.97. It is usually sold under refrigerated conditions and is given a shelf life of 20-30 days, although it is believed that the shelf life of the implicated product was much longer than this.

The exact details of the process employed by the manufacturer of the implicated mascarpone are not reported. However, it is evident that most mascarpone production processes are not designed to achieve total destruction of spores of the most heat-resistant strains of *C. botulinum*, namely the proteolytic strains and in particular the type A strains. In fact, it is clear that the usual process of heating the cream to 80-90°C would actually achieve little, if any, reduction in proteolytic spore numbers and neither would heat processes up to 100-110°C.

Even though the product is called an acidified cheese, with a pH value between 5.7 and 6.5, it would be considered by most microbiologists to be a low acid product and, if destined for ambient storage, it would be expected to receive a thermal process capable of destroying spores of *C. botulinum* - see Chapter 3 (a 'botulinum cook' of 121°C for 3 minutes is generally considered to provide a 12 log reduction in *C. botulinum* spores although the dairy industry usually achieves an equivalent process by heating at very high temperatures of 138-142°C for less time, commonly 2-5 seconds, i.e. Ultra High Temperature (UHT) processes). However, as the mascarpone implicated in this outbreak was destined to be sold under refrigerated conditions, it is reasonable for the manufacturer to consider the application of a 12*D* process to be unnecessary as any spores of proteolytic strains of *C. botulinum* would not be capable of growth under controlled refrigeration (<8°C).

For similar reasons it would not have been considered necessary to formulate the product to achieve a pH capable of inhibiting the growth of proteolytic strains of *C. botulinum*, i.e. pH 4.6 or below. The acidification process for mascarpone manufacture is only used to achieve coagulation of the milk proteins and, although the pH is decreased slightly, this is insufficient to prevent the growth of *C. botulinum* under conditions of temperature abuse. *C. botulinum* type A is generally capable of growth at pH values above 4.6 (Table 3.6) and it is not possible to manufacture a traditional mascarpone cheese with levels of acidity that would be required to decrease the pH to such a low level. As *C. botulinum* type A and other proteolytic strains cannot grow at temperatures below 10°C, it is noteworthy that, as designed, the product was inherently safe with regard to these strains of *C. botulinum*. The outbreak could not have occurred without temperatures exceeding 10°C for sufficient time to allow high levels of the organism to develop and produce toxin. Clearly, under conditions of temperature abuse, the physico-chemical properties of the product could not prevent or significantly inhibit the growth of *C. botulinum* type A.

C. botulinum will be an occasional contaminant of raw milk. Spores present in the soil may enter the raw milk via contamination on the external surfaces of the udder or teats or from contaminated milking equipment. It is not possible to eliminate the organism from raw milk and, although levels can be reduced by effective milking parlour hygiene and cleaning practices, the presence of low levels of spores will occur from time to time. It should be assumed that the organism will be present in milk at a low frequency and at low levels and therefore will also be present in the final product from time to time. Indeed, following this outbreak of botulism, Franciosa *et al.* (1999) assessed the incidence of *C. botulinum* in 1017 samples of mascarpone cheese (878 of which were from 34 different lots produced over a 4 month period by the manufacturer implicated in the outbreak) and found *C. botulinum* spores in 331 samples; 327 of which were from the plant implicated in the outbreak with 325 samples contaminated with *C. botulinum* type A and 2 samples with type B (proteolytic). Four out of 139 samples of mascarpone cheese manufactured by different producers were also found to be contaminated with *C. botulinum* type A at levels of < 10 spores/g.

As the process for making mascarpone will not eliminate the organism and the product, if temperature abused, is capable of supporting the organism's growth, the question arises as to where temperature abuse might have occurred to allow the outbreak to happen.

The first significant point offering opportunity for growth is after the heat process and during the cooling of the mascarpone. Mascarpone is often hot filled into large containers (up to 3 kg) at temperatures as high as 80°C, although it is reported that the weight of the containers implicated in this outbreak was 500 g (Food and Drug Administration, 1996). Depending on the method used to cool the containers, it is possible that cooling to < 10°C, the critical level to prevent growth of proteolytic *C. botulinum*, could have taken many hours. In this circumstance, conditions could certainly allow germination of spores and growth to occur, although the extent to which this occurred would be dependent on the exact conditions of the process. It is reported that typical mascarpone production processes would involve cooling from 70°C–75°C to 36°C in 2 hours and then from 36°C to 4°C in 4 hours (Franciosa *et al.*, 1999) which, if adhered to, would achieve effective control of *C. botulinum*.

Bulk and pre-pack containers should be cooled rapidly after filling to prevent the germination and growth of surviving spore-forming bacteria. The critical temperatures to avoid are, not surprisingly, those over which germination and growth may be fastest, namely between 10 and 45°C. At elevated temperatures, germination and subsequent growth could occur rapidly. It is therefore common practice to chill products from 45°C to < 10°C within 3-4 hours, although many manufacturers aim to chill to < 5°C within 4 hours from completion of heating to prevent germination and growth of pathogens including other clostridia, e.g. *C. perfringens.*

Products such as mascarpone, which are pasteurised and then hot filled, effectively select spore-bearing microorganisms as these will be the only ones capable of surviving the process. If the product is then cooled slowly, these surviving organisms are exposed to conditions of unchallenged growth because of the absence of any viable competitive microflora.

In addition to the possibility of temperature abuse during cooling, it is equally possible that temperature abuse may have occurred during distribution, in retail cabinets or in the home. The equipment and systems of temperature control in distribution and retail display should equally ensure that the 'chill chain' is maintained and, although some refrigeration systems may be tested to their limits to maintain < 5°C during warm summer months, most systems should be capable of preventing temperatures from exceeding 8°C.

It is reported that mascarpone cheese manufactured by the implicated producer was routinely marked 'Keep the product at + 4°C' and clearly, it is critical for products which rely exclusively on refrigerated storage for their safety to have clear labelling evident to all those in the food chain throughout distribution to the consumer.

In addition to the lack of inhibitory factors in this product, it is important to recognise that the product was also allocated an extremely long shelf life. If kept properly refrigerated, *C. botulinum* type A would not have represented a hazard even with such a long shelf life. However, the inherent problem with allocating long shelf lives to products reliant on refrigeration alone for inhibiting multiplication of *C. botulinum* is that they are more vulnerable to temperature abuse due to the number of times they may be removed from the refrigerator, used and then replaced. In addition, long shelf lives may give the customer (and possibly some vendors) the perception of stability and this could lead to complacency with regard to its storage and handling.

In assessing the potential for growth of *C. botulinum* in mascarpone under conditions of temperature abuse Franciosa *et al.* (1999) found that product naturally contaminated with type A spores at < 10/g produced toxin in 3 days at 28°C but did not produce toxin after 20 days at 15°C. When artificially inoculated with type B spores at 100/g, toxin was produced after 34 days at 15°C and 3 days at 28°C.

It is interesting to note that even though the mascarpone, if held under proper refrigerated conditions, would not have allowed the growth of proteolytic strains of *C. botulinum*, the formulation and process did represent a potential hazard with regard to psychrotrophic strains of non-proteolytic *C. botulinum*. These strains could readily survive a process at 80-90°C and have the potential to grow under refrigeration conditions (< 5°C) in the mascarpone product of pH 5.7-6.5 and a_w > 0.97. Such an extensive shelf life makes the hazard in relation to non-proteolytic strains very real indeed.

Approaches to avoiding this outbreak involve improved product formulation and process control during manufacture which, in addition to reducing product shelf life, could have limited the extent of growth of the organism. In addition, effective temperature control at all stages including manufacture, distribution, retail storage and display or with the consumer could have prevented growth of proteolytic *C. botulinum* altogether. Indeed, the fact that all eight patients reported that the mascarpone was malodorous (smelled off) indicated that the hazard existed and, if the product was discarded, the problem would again have been avoided. Any food indicating obvious signs of abnormality should not be consumed.

It is imperative that potential hazards like *C. botulinum* are considered at all stages of the process from raw materials, through all processing steps, final product storage, distribution, retail display and with the consumer, employing a HACCP-based approach. This will help to ensure suitable controls and information are in place allowing all parties to properly discharge their food safety responsibilities and thereby avoid this type of outbreak.

HAZELNUT YOGHURT: UK

The largest recorded outbreak of botulism in the UK occurred in 1989 in which a total of 27 individuals suffered varying symptoms of botulism (O'Mahony *et al.*, 1990). The ages of those affected ranged from 14 months to 74 years with an average age of 29 years. All but one patient were hospitalised and almost half were admitted to intensive care units. One elderly patient, aged 74 years, died due to aspiration pneumonia (Critchley *et al.*, 1989).

The time between consumption of the implicated product and onset of symptoms ranged from two hours to five days although the median onset time was one day. These differences probably related to the significant variation in the quantity of yoghurt consumed by the affected persons (a few spoonfuls to three cartons). The product implicated in the outbreak was hazelnut yoghurt (Table 2.2). Of the 27 affected individuals, 25 confirmed that they had eaten the same product supplied by one manufacturer; the other two had consumed hazelnut yoghurt from a different manufacturer, although this manufacturer had historically used the same source of hazelnut conserve as the other implicated manufacturer.

The organism responsible for the outbreak was *C. botulinum* type B and the toxin was detected in a sealed can of hazelnut conserve (used in the manufacture of the yoghurt) obtained from the premises of the yoghurt manufacturer; the can showed evidence of blowing i.e. the can was swollen. In addition to the detection of the toxin in the conserve, two opened cartons of hazelnut yoghurt retrieved from patients' homes together with 15 unopened cartons of hazelnut yoghurt were also found to contain type B toxin by the mouse bioassay.

The amount of type B toxin in the hazelnut conserve and the contaminated yoghurt was found to be 600–1800 mouse lethal doses per ml (MLD/ml) and 14–30 MLD/ml, respectively. From this data it was estimated that 1750–3750 MLD type B toxin was present per 125 g carton of yoghurt consumed in the outbreak. Samples of opened and unopened hazelnut yoghurt yielded *C. botulinum* type B (proteolytic).

The hazelnut yoghurt was manufactured using a standard process for the production of these types of fermented milk. A mixture of pasteurised milk, milk powder and starch was heated to 82°C, sugar was added and the mixture held at this temperature for 30 minutes prior to cooling, then pumping to an inoculation tank where starter cultures were added prior to being poured into churns. The mixture was fermented for 2–3 hours in

Table 2.2   Outbreak overview: hazelnut yoghurt

Product type:	Yoghurt flavoured with hazelnut conserve
Year:	1989
Country:	UK
Levels:	Not reported (toxin and organism detected in conserve and yoghurt)
Cases:	27 (1 death)
Type:	*C. botulinum* type B (proteolytic)

Possible reasons

Conserve manufacturer
(i)   Substitution of sweetener used which raised the water activity of the conserve from 0.90 to 0.99; within growth range of *C. botulinum*
(ii) 'Botulinum cook' not applied to low sugar, ambient stored, canned product
(iii) Blown cans of low sugar conserve were not recognised as signs of instability in relation to hazard of *C. botulinum*

Yoghurt manufacturer
(iv) Use of canned, ambient stored hazelnut conserve (low sugar/low acid) not subjected to a 'botulinum cook'

Control options*
(i)   Conserve processed to ensure an equivalent 'botulinum cook' (121°C, 3 minutes)
(ii) Conserve formulated to prevent growth of *C. botulinum* (pH 4.6 or less, a_w 0.94 or less)
(iii) Conserve stored chilled at < 8°C
(iv) Investigation of conserve spoilage to assess reasons and broader implications
(v)   Yoghurt manufacturer sourcing policy to assure safety of incoming raw materials

*Suggested controls are for guidance only and may not be appropriate for individual circumstances. It is recommended that proper hazard analysis is carried out for every process and product to identify where controls must be implemented to minimise the hazard from *C. botulinum*.

ten-gallon churns in an incubator. The final product was made by mixing hazelnut conserve with the yoghurt base prior to dispensing each churn's contents into 360 cartons. The product was allocated a total shelf life of 27 days. No evidence of inadequate hygiene practices or production methods at the dairy were reported. The product was labelled with advice to keep the yoghurt refrigerated and to consume it within two days of purchase (O'Mahony *et al.*, 1990).

The ingredient responsible for this botulism outbreak was unquestionably the hazelnut conserve added to the yoghurt.

The hazelnut conserve was manufactured using roasted hazelnuts, water, starch and other ingredients by preheating these to 90°C for 10 minutes prior to filling the mixture into cans. The sealed cans were retorted in boiling water for a minimum of 20 minutes and apparently sold as an ambient stable, canned conserve (O'Mahony *et al.*, 1990).

The conserve was primarily manufactured using sugar as the sweetening agent. However, it is apparent that in July 1988, 11 months before the outbreak, 76 cans were manufactured using an artificial sweetener, aspartame, in place of sugar and 36 of these cans were supplied to the manufacturer of the hazelnut yoghurt implicated in 25 of the botulism cases. Some of the conserve had also been supplied to other manufacturers of hazelnut yoghurt, including the manufacturer implicated in the two other cases involved in this outbreak. It was reported that the latter manufacturer claimed to have used this conserve many months before this incident (O'Mahony *et al.*, 1990).

It is of interest to note that some companies who had received cans of the aspartame sweetened conserve had reported spoilage evidenced by blowing of cans and, in October 1988, the conserve manufacturer added potassium sorbate to the conserve to control what was assumed to be yeast spoilage.

The water activity, a_w, of the hazelnut conserve sweetened with sugar was reported to be 0.90 but, after substitution with aspartame, this increased to 0.99 (Rhodehamel *et al.*, 1992). Although the pH of the conserve is not reported, 17 of the conserve cans were tested for pH during the outbreak and 15 out of 17 had a pH value between 5.0 and 5.5, with the other two having pH values of 4.5 and 4.7.

Most food technologists posed with the question of which food types are least likely to be associated with a botulism outbreak would have included yoghurt fairly high on the list – prior to this outbreak! In fact, the conditions in place at the manufacturer of the yoghurt were apparently well controlled and capable of delivering a safe 'white base', a term often applied to the base yoghurt prior to the addition of flavouring. However, what is clearly demonstrated in this outbreak is the requirement for a sound raw material sourcing policy. The yoghurt manufacturer should reasonably expect the supplying conserve manufacturer to ensure the safety of the raw material but should have ascertained the critical safety information for each batch of product received, e.g. process conditions, pH, water activity.

The manufacturing process of the hazelnut conserve implicated in this outbreak is similar to that of many canned, shelf-stable products. The raw material mixture is subjected to a heat process to destroy contaminating microorganisms in a can which, together with effective sealing and control of post-process contamination, should have ensured a safe, stable product. However, safety and stability can only be achieved if either the heat process is sufficiently high to destroy all vegetative and spore-forming pathogenic and spoilage microorganisms or the product formulation is capable of preventing growth of any surviving organisms remaining in the can.

It is generally recognised that to destroy spores of *C. botulinum* under otherwise optimal conditions requires a heat process of at least $121°C$ for 3 minutes, the $12D$ 'botulinum cook' (see Chapter 3). To achieve an equivalent thermal process in this product at $100°C$ would require a process lasting 600 minutes (based on D_{100} of 50 minutes, International Commission on Microbiological Specifications for Foods, 1980). It is therefore easy to see that a 20-minute heat process given to the cans would have been insufficient to achieve even a 1 log reduction in *C. botulinum*.

The fact that the hazelnut conserve preserved with sugar did not previously result in an outbreak of botulism is attributable to the original sugar based formulation. Sugar is an effective humectant, decreasing the water activity of the product. The reported water activity of the sugar-preserved conserve was 0.90 which is significantly below the growth-controlling water activity, a_w, of 0.94 for *C. botulinum* (see Chapter 3). Therefore, the original sugar formulated conserve would have been safe with regard to the hazard of *C. botulinum,* even under ambient storage. However, upon substitution of the sugar with aspartame, the conserve was left without a key botulinum controlling factor. Aspartame is used in much less quantity than sucrose to achieve the required sweetness and has almost negligible effect on water activity, as can be seen by the resulting a_w of the conserve which increased from 0.90 with sugar to 0.99 with aspartame. The pH of the conserve was not low enough to prevent growth of *C. botulinum* and the heat process did not make the conserve ambient stable. If spores of the organism were present in the raw materials then it is clear that the aspartame-containing conserve would have allowed surviving spores to grow.

Of course, the other clear warning sign in this incident was provided by reports to the conserve manufacturer of 'blowing cans' following supply of the conserve containing aspartame. It is not reported whether the nature of the can spoilage was investigated by the conserve manufacturer or whether it was just assumed to be yeast spoilage. However, it is critical with any spoilage of a canned product to determine whether the responsible microorganisms survived the process or were introduced as post-process contaminants. Simple microbiological tests using direct microscopy or culture on selective and/or non-selective agar (incubated aerobically and anaerobically) followed by microscopy, e.g. Gram stain, would suffice.

If the organism is a vegetative contaminant, such as yeast, then it is most likely to be caused by post-process contamination, whereas if it is a bacterial sporeformer, it should raise the suspicion of process survival and provide some warning about the sufficiency of the process and stability of the final product. Other possible options available to make this product safe would have been to distribute and store the canned product under refrigeration or to acidify the purée. Such options may not be readily adopted by manufacturers due to effects on cost, convenience and eating quality.

Application of a hazard analysis approach to the production of hazelnut conserve would have resulted in the identification of a clear hazard relating to the reformulated aspartame-containing hazelnut conserve and the consequent need to modify the conserve processing conditions to destroy *C. botulinum* or modify the formulation of the product to control its growth during storage.

GARLIC IN OIL: CANADA AND USA

A large outbreak of botulism occurred in Canada in 1985. A total of 36 people suffered botulism linked to the consumption of a food at a restaurant in Vancouver, British Columbia (Blatherwick *et al.*, 1985a, Blatherwick *et al.*, 1985b, St Louis *et al.*, 1988). The first recognised cases were two sisters and their mother who had eaten at the restaurant and the subsequent cases were only identified after a general alert was raised. It transpired that all of those affected had eaten at the implicated restaurant, although the outbreak occurred in two clusters between July and September 1985. Due to the variety of atypical symptoms and the wide dispersal of affected individuals, most cases were originally misdiagnosed; although seven persons required mechanical ventilation (Dodds, 1990), no deaths were reported. The causative organism was *C. botulinum* type B which was recovered from the faeces of one patient and type B toxin was found in the serum of three patients.

The food implicated in this outbreak was chopped garlic in soybean oil (Table 2.3) which was used to prepare garlic butter that was subsequently spread on sandwiches. The garlic in oil was a commercial product prepared in the USA from sun-dried, chopped garlic which was re-hydrated and placed in soybean oil. No further process was given and no preservative factors were added to the garlic product prior to sale, although the product label contained instructions to refrigerate (St Louis *et al.*, 1988).

Another outbreak of botulism implicating a similar product, chopped garlic in olive oil, was reported to have occurred in the USA in February 1989. Three people suffered symptoms of botulism in Kingston, New York, and the causative organism was *C. botulinum* type A (Morse *et al.*, 1990, Anon, 1989). All individuals were hospitalised, although no deaths occurred. Pre-formed *C. botulinum* toxin type A was detected in samples of the prepared chopped garlic in olive oil. The product was used in a spread for garlic bread at a dinner party. The pH of the garlic in oil was reportedly 5.7 and the measured water activity 0.932, although the validity of a water activity measurement of such an oil based product is subject to question. The product was manufactured between 1985 and September 1987 by mixing garlic, ice water and olive oil, without the addition of any chemical or acidification agents. Although labelled with an instruction to keep refrigerated, in small print, one patient recalled receiving a jar as a gift in the summer of 1988, storing it at room temperature for three months and then, after opening, storing it in the refrigerator and using it over a subsequent six-month period.

Table 2.3   Outbreak overview: garlic in oil

Product type:	Chopped garlic in oil (soybean oil and olive oil)
Year:	1985 and 1989
Country:	Canada and USA
Levels:	Levels not reported
Cases:	36 (Canada) and 3 (USA) (no deaths)
Type:	*C. botulinum* type B (Canada) and type A (USA)

Possible reasons

(i)   No thermal destruction process applied to garlic to effectively destroy spores
(ii) No controlling factors added to garlic in oil, i.e. acid or salt
(iii) Anaerobic storage of garlic (in oil)
(iv) Extended life given to product
(v)   Temperature abuse during storage of product

Control options*

(i)   Heat process garlic to destroy spores of *C. botulinum* (121°C, 3 minutes)
(ii) Acidification of garlic to pH 4.6 or less or reduction in water activity by use of salt to achieve an aqueous salt level of 10% or greater
(iii) Effective refrigeration and temperature control
(iv) Reduce shelf life to limit opportunities for long period of temperature abuse

*Suggested controls are for guidance only and may not be appropriate for individual circumstances. It is recommended that proper hazard analysis is carried out for every process and product to identify where controls must be implemented to minimise the hazard from *C. botulinum*.

Both outbreaks are believed to have occurred due to elevated temperature storage of the garlic products allowing germination and growth of contaminating *C. botulinum* spores to toxin-forming levels. In the first outbreak it is reported that the garlic in oil had been held at ambient temperatures for several months at the restaurant concerned.

Garlic in oil products are manufactured in a variety of ways usually involving chopping fresh garlic and mixing with oil, although in one of the outbreaks the use of sun-dried then re-hydrated garlic is reported. The mixture of garlic in oil is usually stored in glass jars/bottles. The products implicated in these two outbreaks were not subject to any form of heat or chemical preservation process, except that both were reported to be intended for refrigerated storage.

Fresh garlic has a high water activity (> 0.97) and slicing/chopping the cloves would undoubtedly serve to both release nutrients from the garlic flesh and transfer any contaminating *C. botulinum* spores from the waxy intact outer surfaces to the nutritious, exposed cut areas. Garlic is a root vegetable and although protected by layers of dry leaves, the outer surfaces are likely to carry microorganisms from the earth in which the garlic is grown.

C. botulinum is frequently isolated from soil samples in surveys although the incidence, levels and type vary depending on the location and climate (Table 1.5). *C. botulinum* must be considered to be a potential contaminant of any root vegetable from time to time and should be taken into account in any assessment of risk associated with their subsequent processing and storage.

Vegetables such as garlic are often stored in oil due to the preservative effect that anaerobic conditions have on preventing spoilage by aerobic microorganisms. Some manufacturers bake garlic prior to adding it to oil to destroy vegetative spoilage microorganisms, such as yeasts. However, baking may not destroy some bacterial spores and, clearly, the storage of chopped garlic under anaerobic conditions is likely to allow the growth of *C. botulinum* if other conditions are not adjusted to be inhibitory to its growth. The effective preservation in relation to most spoilage microorganisms, conferred by storage of the garlic in oil, ultimately contributed to some of these botulism outbreaks as even under conditions of temperature abuse the product rarely succumbed to spoilage by other microorganisms. Otherwise the product would, presumably, have been inedible, as in the case described by St Louis *et al.* (1988) in which the product was eventually discarded because it became malodorous.

Although there have been some reports that allicin is a natural inhibitory (antimicrobial) compound present in fresh garlic, it is clear from these outbreaks that this cannot be relied upon to prevent the growth of *C. botulinum*. It is important to recognise that this product would have been safe with regard to proteolytic strains of *C. botulinum* if refrigeration conditions had been effectively applied and maintained. It must be noted however, that even under refrigeration, spores of non-proteolytic strains of *C. botulinum* capable of psychrotrophic growth would still present a hazard and, as such, the process and product should certainly have been designed and formulated to, at least, control these strains. Alternatively, the refrigerated shelf life could have been limited to prevent such strains from growing to dangerous levels.

In order to have prevented germination and growth of proteolytic strains of *C. botulinum* in the implicated product, other than effective refrigeration throughout the life of the product, three control options exist; heat processing, product formulation and shelf life conditions.

Although application of a heat process sufficient to achieve an acceptable reduction in contaminating spore numbers (121°C for 3 minutes) would have made the product safe, it is unlikely that such processing would give an organoleptically acceptable product. The most appropriate option for manufacture of a safe product would be to formulate the product to prevent growth of *C. botulinum* during the assigned shelf life. Garlic could be chopped and mixed with a suitable acid to achieve an effective means of preservation providing the pH was reduced to pH 4.6 or below. It is important to recognise that such treatment may need to be combined with a mild heat process to destroy any vegetative contaminants left in the garlic which may be capable of growth and spoilage of the product or causing localised elevation of the pH to within the growth range of *C. botulinum*. An alternative approach could have been to reduce the water activity of the garlic by the addition of a humectant such as sodium chloride to achieve an aqueous salt content of 10% or greater; this is widely recognised as the level sufficient to prevent the growth of proteolytic strains of *C. botulinum* (Table 3.6).

The option chosen would depend on the intended use of the garlic in oil, as a high acidity or high salt content would need to be compatible with the flavour requirement of the final product. In many cases, garlic is often present in oil to flavour it so a high salt or acid level has little consequence for the oil phase itself. Another, and perhaps the most simple, approach that could have yielded a safe product would have been to substantially reduce the refrigerated shelf life of the garlic in oil. Alternatively, the restaurant could have made the garlic in oil on a daily basis as a fresh product. This is clearly a case where a fresh product would have been concomitant with a safer product.

It is of some concern that products, such as garlic in oil, sold under ambient conditions and with shelf lives in excess of 12 months are becoming an increasingly familiar sight in the UK and worldwide market. Although the outbreaks discussed have involved manufactured products, many incidents of botulism are associated with home-produced products as seen in the recent UK outbreak caused by home-preserved mushrooms in oil from Italy (Anon, 1998b). All of these outbreaks are clear reminders of the danger posed by these products if the formulations, process and/or storage conditions are not under careful control.

Following the outbreak caused by garlic in oil in the US, action was taken by the US and Canadian authorities that required commercial garlic in oil products to be formulated to prevent the growth of *C. botulinum* (Anon, 1989). Reliance on temperature control alone is not considered by these authorities to provide sufficient safety for the product and it is recommended that the product should be both acidified to a pH of 4.6 or below and stored under refrigeration.

The lessons from these outbreaks have clearly not been learnt by many manufacturers and retailers of these products, possibly because many are manufactured in cottage industries with little technical resource capable of identifying the hazards and determining appropriate controls. It is perhaps more surprising that national/government health departments have not yet acted in a manner similar to the responsible position taken by the US Department of Health and Human Services in providing strong guidance on the safe manufacture of these products. Although government health departments may be concerned about a backlash from supporters of such small industries, it is surely appropriate to provide precautionary advice before similar outbreaks occur in other countries.

BAKED POTATOES: USA

At least five incidents have occurred in which baked potatoes have been implicated as the causative agent of botulism (Table 2.4). In 1978, seven cases of botulism were recorded (Seals *et al.*, 1981) following the consumption of a potato salad in a restaurant in Denver, Colorado. No deaths occurred. All of those affected had eaten potato salad and the causative organism was identified as *C. botulinum* type A. In 1984, one case of botulism (*C. botulinum* type A) was reported following the consumption of a baked potato from a restaurant in Alabama (MacDonald *et al.*, 1986) and in 1992, a home associated outbreak due to potato salad resulted in two cases of botulism again, in Denver, Colorado (Brent *et al.*, 1995). In this last outbreak, both the toxin and spores of *C. botulinum* type A were detected in the stool samples of one of the patients and the leftover potato salad. Although not entirely conclusive, the largest outbreak implicating potatoes occurred in Clovis, New Mexico, in 1978 (Ryan *et al.*, 1978; Mann *et al.*, 1983). A total of 34 cases of botulism were recorded with all patients being hospitalised; two patients died. The outbreak strain was *C. botulinum* type A and the toxin was isolated from clinical samples and from a potato salad from a restaurant salad bar that all patients had visited. In 1994, another large outbreak of botulism occurred in El Paso, USA,

Table 2.4   Outbreak overview: baked potatoes

Product type:	Baked potato or baked potatoes used for potato salad
Year:	1978 ($\times$ 2), 1984, 1992 and 1994
Country:	USA
Levels:	Levels not reported
Cases:	74 (total) (2 deaths)
Type:	*C. botulinum* type A

Possible reasons
(i)   Ambient storage of potatoes (some foil wrapped) for several days after baking
(ii)   Use of temperature-abused, cooked potatoes in ready-to-eat product

Control options*
(i)   Chilling of baked potato
(ii)   Use of baked potato on day of bake only
(iii) Use of baked potato in potato salad immediately and chilled storage of pH controlled salad for longer periods

*Suggested controls are for guidance only and may not be appropriate for individual circumstances. It is recommended that proper hazard analysis is carried out for every process and product to identify where controls must be implemented to minimise the hazard from *C. botulinum*.

affecting 30 people who ate dips prepared with baked potato at a Greek restaurant (Angulo *et al.,* 1998). Type A botulinum toxin was detected in samples from patients and in the dips.

Although the exact details are not recorded for all of the outbreaks, it is clear from both the restaurant and home-associated outbreaks in Colorado and the recent outbreak in El Paso that the baked potato and/or the potato salad/dip made from it were subjected to extensive temperature abuse. The baked potatoes used in the restaurant outbreak in Colorado were unwashed, wrapped in aluminium foil and then baked at 260°C for 45 minutes. They were generally used on the same day but any leftover potatoes were kept at ambient temperatures for up to five days prior to being peeled and used in the potato salad. Similarly, in the El Paso outbreak, the potatoes were foil wrapped and baked at 250°C for approximately two hours and then reportedly kept at room temperature for 18 hours prior to use. It is however believed that the potatoes may have been kept for longer periods (Angulo *et al.*, 1998). The pH of the potato salad in the Colorado incident in 1992 (home-associated) was reported to be 4.85 and from the El Paso incident 3.7.

It is easy to see how outbreaks of this nature can occur; it is equally easy to see how they can be prevented. *C. botulinum* spores are likely to be an occasional contaminant on raw root vegetables which grow in the earth and, when harvested, usually retain some soil. Surveys of the incidence of *C. botulinum* spores in soil have readily found them to be present although the incidence and levels vary, as does the type of *C. botulinum* (Table 1.5).

Baked potatoes, produced either in a home setting, restaurant/catering or under industrial manufacturing conditions may be seen by many people to be fairly innocuous in relation to the hazard of botulism. However, spores of *C. botulinum* will occur on potatoes and as the product can also support their growth and toxin production if the spores gain entry to the flesh (Sugiyama *et al.*, 1981), it must be considered a potential hazard to the product. This is particularly relevant as it is common practice to pierce potatoes prior to baking to prevent excess pressure building up and the potatoes bursting.

Under the conditions of baking a potato, although oven temperatures may exceed 200°C for over an hour, the potato flesh itself will rarely achieve temperatures in excess of 100°C (Seals *et al.*, 1981, Sugiyama *et al.*, 1981). Clearly, the application of a full 'botulinum cook' in the preparation of a baked potato, while it would undoubtedly make it safe, is not a viable option in practice. Wrapping potatoes in foil may have contributed to the anaerobic conditions that facilitated the growth of *C. botulinum* subsequent to cooking, but it is important to recognise that even without foil, conditions several millimetres beneath the surface of the potato tuber will not necessarily be aerobic and non-wrapped potatoes may be equally vulnerable to growth where spores penetrate deeper into the tuber.

Sugiyama *et al.* (1981) demonstrated that when potato tubers were surface and stab inoculated prior to baking in foil, time to toxin formation by a *C. botulinum* type A strain was inversely related to spore dose. Potatoes became toxic within 3–4 days when held at 22°C or 30°C after inoculation with 10^5 spores and baked at 204°C for 50 minutes, whereas toxin detection only occurred after 6–7 days when inoculated with 10 spores. In general, stab inoculation resulted in slightly shorter times to production of toxin than surface inoculation (Table 2.5).

Table 2.5   Production of toxin by *C. botulinum* type A in foil wrapped baked potatoes*, adapted from Sugiyama *et al.* (1981)

Inoculum level (spores)	Surface inoculation (days to toxin formation)		Stab inoculation (days to toxin formation)	
	Storage at 22°C	Storage at 30°C	Storage at 22°C	Storage at 30°C
10^5	4	4	3	3
10^3	5	5	4	3
10^1	7	6	7	6

*Baked at 204°C for 50 minutes after inoculation.

It is of little doubt that these reported outbreaks were caused by the extended storage of potatoes at ambient temperatures after baking. Extended warm storage of baked potatoes provides ideal conditions for the growth of surviving heat-resistant microorganisms. Spores of clostridia and *Bacillus* spp. would be favoured under these conditions as any vegetative competitive microorganisms would be destroyed by the baking process leaving the surviving spores with every opportunity to germinate and grow.

Interestingly, in the studies by Sugiyama *et al.* (1981) it was shown that while *C. botulinum* type A produced toxin in baked potatoes within three days under ambient storage, no toxin was produced in raw potato, stab inoculated with 10^5 spores, then foil wrapped and stored at 30°C for seven days.

To make the baked potato safe for consumption would have necessitated either holding the potato hot, i.e. >60°C, after baking, cooling and then storing under refrigerated conditions, i.e. <8°C, or, if stored at ambient temperature, used on the day of baking. Although keeping the product hot could work for short periods, excessive time at high temperatures could result in significant deterioration in product quality. Keeping the product chilled would certainly have made it safe with regard to *C. botulinum* type A and would have allowed a safe shelf life up to the five days used by the restaurant.

As the baked potato was destined to be made into potato salad, an alternative option would have been to make the potato salad on the same day as the potato was baked and allow the low pH mayonnaise to confer some degree of stability on the potato. However, in order to make it safe for storage at ambient temperature, the potato would have needed to be acidified to a pH of 4.6 or less and, as this would also require equilibration of pH throughout the potato, safe production of this in a restaurant or in the home is unlikely to be achievable and storage of freshly made potato salad would still have required refrigeration.

As with many of the preventative measures for controlling outbreaks, it is often the simplest rules that make most sense. In this case the normal rules given to the public regarding 'leftovers', if followed, could have prevented this outbreak, i.e. keep leftover foods hot or cool them down and keep them cold. Indeed, the United States Food and Drug Administration advocates that cooked, potentially hazardous foods should be kept at temperatures of 5°C or less or at 60°C or above (Food and Drug Administration, 1999), advice which needs to be heeded, especially by those involved in preparing such products commercially.

BOTTLED PEANUTS: TAIWAN

An outbreak of botulism occurred in 1986 in Taiwan. A total of nine cases were reported, seven of which were employees of a printing factory who had all consumed the implicated product in the factory canteen 2–3 days prior to the onset of symptoms (Chou *et al.*, 1988). Four of the employees were hospitalised for up to 105 days and two required mechanical ventilation for a prolonged period. One of the employees, the cook, died due to respiratory complications three weeks after admission to hospital. Two other individuals, a grandmother and her grandson were later also found to have suffered botulism associated with the consumption of the same implicated product. The woman, who was 68 years old, suffered symptoms 24 hours after consumption of the product and died within 24 hours. The product implicated in the outbreak was commercially bottled peanuts produced by a company in southern Taiwan (Table 2.6).

The organism responsible was *C. botulinum* type A and type A toxin was detected at a dilution of 1:10 000 in one of four unopened jars of peanuts retrieved from the printing factory canteen. In addition, the opened jar of

Table 2.6 Outbreak overview: bottled peanuts

Product type:	Commercially heat processed bottled peanuts in liquid (water/sauce)	
Year:	1986	
Country:	Taiwan	
Levels:	Levels not reported	
Cases:	9 (2 deaths)	
Type:	*C. botulinum* type A	

Possible reasons
(i)   Contamination of shelled peanuts with spores of *C. botulinum*
(ii) Heat process unable to achieve 12 log reduction in *C. botulinum* spores
(iii) Peanuts bottled in liquid without any botulinum controlling factors
(iv) Ambient storage for extended period

Control options*
(i)   Effective washing of peanuts to reduce spore loading
(ii) Retorting of product to achieve $F_0 3$ or equivalent by boiling for longer periods
(iii) Storage of peanuts in brine solution of 10% or greater aqueous salt or acidification to pH 4.6 or less

*Suggested controls are for guidance only and may not be appropriate for individual circumstances. It is recommended that proper hazard analysis is carried out for every process and product to identify where controls must be implemented to minimise the hazard from *C. botulinum*.

peanuts recovered from the grandmother's home was also found to contain type A botulinum toxin at the same dilution.

The product implicated in all cases came from the same batch code of peanuts and it is pertinent to note that the product consumed by the latter two cases was purchased over one month after the issue of a public recall notice by the local Department of Health following the identification of the outbreak from the first seven cases in the printing factory.

A total of 104 jars of the implicated batch of unopened peanuts were retrieved from local stores and of these 34 (33%) were found to contain type A botulinum toxin. This contrasted with negative results from tests carried out on 32 jars of other batches of the same product.

The company responsible for manufacturing the product implicated in this outbreak was a small family owned business which was reported not to hold a licence for the manufacture of canned foods. The product was manufactured without the use of a retort or steam pressurised equipment, using dry shelled peanuts from a local source in Taiwan. The peanuts were rinsed in water and then boiled in an open cauldron for three hours. They were left to cool and then the peanut/water mixture was hand filled, using spoons, into steam cleaned glass jars. Lids were manually placed and tightened. The jars were then placed in another open cauldron and steamed for an hour, after which they were allowed to cool to ambient temperature. The product was produced in batches of 500 kg quantities to make approximately 1000 × 250 g jars and 600 × 400 g jars which were supplied to local distributors who in turn supplied the product to retail outlets. The manufacturer had been producing this product for over six years and reported that some jars developed bubbles or other signs of spoilage which required disposal (Chou *et al.*, 1988). The implicated product was manufactured only 2–3 weeks before the first cases occurred.

This outbreak was unquestionably caused by a combination of under-processing and lack of adequate preservative factors of a food commodity susceptible to both contamination by and growth of *C. botulinum*.

Peanuts are certainly an unusual vehicle for a botulism outbreak but clearly, as ground nuts, they have the potential to be contaminated with spores of the organism from the soil which may be transferred to nuts during shelling. Although washing of the nuts was reported to have taken place, this process step is not likely to remove spores completely. Given the marginal nature of the heat process it is certainly possible that a moderate spore loading on the peanuts, if not washed or insufficiently washed, could result in an increased possibility of process survivors. This is reinforced by the fact that such a large outbreak occurred from one single batch after over six years of processing the same product. This could indicate that some factor may have changed in relation to that batch which may have led to greater potential for *C. botulinum* contamination, survival or growth. Indeed, it has been postulated that a batch of nuts may not have been rinsed or may have received less than normal processing during cooking (Chou *et al.*, 1988).

Because of the information that signs of spoilage were observed in previous product batches, Chou *et al.* (1988) comment that this was unlikely to have been an isolated incident, presumably suggesting unrecognised illnesses associated with consumption of these products in the preceding years.

Fresh peanuts, having a moisture content varying from 40 to 60%, would generally not support the growth of many bacteria. They are, of course, prone to mould growth and are therefore usually dried to moisture levels below 10%. However, as they have a neutral pH, the water/peanut mixture would provide conditions either within the nut or at the interface between the nut and the liquid sufficiently conducive to growth and toxin production by *C. botulinum*. Peanuts contain high quantities of fat together with some protein and they can therefore provide ample nutrients for the growth of microbial contaminants if the water activity is sufficiently high, as would have been the case in this product.

Two main practical options exist for the control of *C. botulinum* in this product; firstly the product could be processed to destroy the spores of *C. botulinum* by the application of a 'botulinum cook'. The process applied in this outbreak would, at best, have exposed spores to a combined heat process of 100°C for 3-4 hours giving approximately a 3-5 log reduction in spores of *C. botulinum* based on a D_{100} of 50 min (International Commission on Microbiological Specifications for Foods, 1980). However, given the fact that the peanuts were spooned into jars after the first cook (three hours) and contamination could have been introduced at that point, the true process should actually be considered to be only the final cook of the one hour steaming process which, when taking account of the time required for the product temperature to rise, is likely to have exposed the peanut mixture to temperatures of 100°C for significantly less than one hour, if any reached this temperature at all. This would have achieved at most a 1 log reduction, clearly demonstrating how far below the margins of safety the process could have been. The alternative option to achieve safety would have been by altering the formulation of the product and cooking the peanuts in brine rather than in plain water. The brine concentration required to prevent growth of proteolytic *C. botulinum* is recognised as being 10% or greater salt (sodium chloride) in the aqueous phase of the product and, in combination with the heat process employed in the manufacture of this product, would have rendered it safe and ambient stable.

It is unlikely that, given the fact that the production facility manufacturing the product was unlicensed, the personnel had any specific knowledge of the controls required for the microbiological safety of the product. Safety, in this case, appeared to be based on historical precedence, with an outbreak likely to occur on any occasion where the conditions conspired to allow spores of *C. botulinum* to be present, survive the process, grow and produce toxin in the final product. Indeed, such lack of understanding is undoubtedly commonly associated with the numerous home processes associated with outbreaks of botulism that occur throughout the world involving vegetables. Most people's primary consideration is to ensure that the product is processed sufficiently to prevent overt spoilage. There is little or no appreciation of the hazard created by not achieving temperatures sufficient to destroy spores of *C. botulinum*.

In the outbreak associated with the bottled peanuts, although spoiled products were reported to be discarded, a lack of understanding of the inherent hazard was demonstrated by the failure to recognise the warning signs presented by the spoiled product. Warning signs of this nature should alert manufacturers to the fact that the product may be receiving insufficient processing to destroy microorganisms in the raw material or that contamination is occurring post processing. Alternatively, if the product is designed to achieve stability by formulation, then such signs of spoilage are clear evidence that stability has not been achieved. Spoilage in an ambient stored, heat processed, bottled/canned product should always be considered as a sign of the potential for survival and growth of *C. botulinum* and any resulting investigation must consider the organism and whether the conditions that allowed spoilage may also allow pathogen survival and growth.

CANNED CHEESE SAUCE: USA

In October 1993, eight customers of a local delicatessen in Georgia, USA, suffered symptoms of botulism (Townes *et al.*, 1996). The symptoms were generally mild and included some neurological symptoms together with gastrointestinal illness, with the most common being a dry mouth, difficulty with speaking and difficulty in swallowing. The onset times ranged from 1 to 6 days after consumption of the implicated product and, of those affected, three did not seek immediate medical attention due to the mild nature of the symptoms suffered, whereas five individuals were hospitalised in intensive care after botulism was suspected; one person died. It is interesting to note that of the five individuals hospitalised, four were initially diagnosed with illnesses other than botulism and one individual, who suffered visual problems, was actually prescribed glasses to correct their poor vision! All of those who suffered botulism had consumed the same product purchased from a local delicatessen store on 1 October 1993. In a subsequent case control study, 14 other people who did not suffer botulism consumed foods from the delicatessen on the same day but did not consume the implicated food.

The organism responsible for the outbreak was *C. botulinum* type A and the organism was isolated from the stool samples of four patients. Serum samples taken from two of the patients most seriously affected contained low levels of botulinum toxin.

The implicated product was a baked potato containing cheese sauce (Table 2.7). Cheese sauce left over from the can used for garnishing the potato contained at least 400 mouse lethal doses of type A botulinum toxin per gram and cultures of the sauce also yielded *C. botulinum* type A. A subsequent investigation also found *C. botulinum* type A on the surface of raw potatoes in the delicatessen, from a cutting board used to prepare the potato and the refrigerated display case in which the cheese sauce was stored.

No defect could be found in the manufacturing process of the cheese sauce when an investigation was conducted at the premises of the producer of the product. The cheese sauce was processed at a temperature of 135.5°C for 11.7 seconds and then aseptically filled into cans. No evidence of underprocessing could be found at the site and records indicated that the implicated products were from a batch of nearly two-and-a-half-thousand cans manufactured over one year prior to the incident. No evidence was found of any similar problems associated with cans from the same or other batches of cheese sauce. In addition, two cans retrieved

Table 2.7 Outbreak overview: canned cheese sauce

Product type:	Canned cheese sauce used to garnish a baked potato
Year:	1993
Country:	USA
Levels:	Levels not reported
Cases:	8 (1 death)
Type:	*C. botulinum* type A

Possible reasons

(i) Underprocessing of cheese sauce (not believed to be the case)
(ii) Contamination of cheese sauce with *C. botulinum* after opening can and growth of the organism during opened shelf life
(iii) Inadequate refrigeration of cheese sauce after opening
(iv) Customer instruction to 'Keep refrigerated' after opening insufficiently clear
(v) No controlling factors to prevent growth of *C. botulinum* in cheese sauce after opening the can
(vi) Cheese sauce consumed with no further heat process

Control options*

(i) Effective refrigeration once can is opened
(ii) Clear labelling of product in large and bold print to 'Keep refrigerated'
(iii) Formulation of cheese sauce to prevent growth of *C. botulinum*, e.g. pH 4.6 or less or salt 10% or greater (aqueous phase)
(iv) Limit opened shelf life of the cheese sauce

* Suggested controls are for guidance only and may not be appropriate for individual circumstances. It is recommended that proper hazard analysis is carried out for every process and product to identify where controls must be implemented to minimise the hazard from *C. botulinum*.

from the same batch following the outbreak showed no signs of a process failure. One of the cans was tested for botulinum toxin and proved negative while the other can showed no evidence of swelling after incubation at 35°C for 14 days (Townes *et al.*, 1996).

The baked potato was manufactured in the delicatessen by taking washed, refrigerated potatoes and cooking them on demand in a microwave oven for 6–8 minutes until they were soft. The potato was cut open and then filled with margarine, hot barbecued pork, barbecue sauce and cheese sauce. Cheese sauce was the only ingredient not to be served in other items on the delicatessen's menu. The delicatessen had re-opened approximately eight days prior to this outbreak following a period of extended closure and the proprietor had indicated that the can of cheese sauce opened on that day was stored in a refrigerated cabinet and used in the implicated potatoes on subsequent days. The sauce was decanted daily from the original can into a small plastic cup for use in garnishing the potatoes (Townes *et al.*, 1996).

This outbreak is most likely to have occurred due to contamination of the cheese sauce after the can was opened and subsequent temperature abuse of the product which allowed growth of *C. botulinum* and toxin production. Although the sauce was labelled 'Refrigerate after opening', it is believed that the sauce was inadequately refrigerated after opening and, together with contamination of the sauce with *C. botulinum* spores from potatoes or surfaces/utensils in the delicatessen, this contributed to the outbreak (Townes *et al.*, 1996). It is also of considerable importance to note that while the cheese sauce was apparently processed effectively and labelled appropriately, the lettering on the label was reported to be in extremely small letters of approximately 2 mm height, making the instruction difficult to read.

Clearly, there are a number of lessons that can be learnt from this outbreak. Firstly, the cheese sauce, while being processed safely, was not an ambient stable product once opened. The implicated can of product in this outbreak had a pH of 5.8 and a water activity of 0.96 and the perishability of the product had been recognised by the manufacturer as it had been labelled 'Refrigerate after opening'. However, as this was so critical to the safety of the product in use, it follows that the instructions should have been large, bold and clear, to be seen and be read. In addition, although it is not reported in this case, it is often normal practice for manufacturers of canned products which are perishable after opening to provide some advice on the durability of the product, e.g. 'Refrigerate after opening and consume within three days'. It remains unclear whether the proprietor of the delicatessen had read the label and disregarded this or whether lack of clarity on the label led to improper storage. Indeed, the proprietor had indicated that the cheese sauce was refrigerated after opening but this is in doubt (Townes *et al.*, 1996) as proper refrigerated storage would have precluded the potential for growth and toxin production by this organism.

Challenge test studies conducted on the cheese sauce following the outbreak demonstrated that *C. botulinum* isolated from the implicated cheese sauce and inoculated into fresh sauce at a level of 10^3 spores/20 g sauce, produced toxin after eight days at 22°C, without showing any signs of spoilage. Cheese sauce stored at 5°C for 120 days showed no signs of growth of the organism or toxin production, but when transferred to 22°C storage, growth and toxin production were detected in 15 days.

High water activity cheese products are known to readily support the growth of *C. botulinum* with the most significant recognised hazard being in processed cheese and processed cheese spreads. Although this product was clearly not a processed cheese, the *C. botulinum* controlling factors are likely to be somewhat similar. It has been recognised for many years that processed cheese stability at ambient temperatures with regard to preventing the growth of *C. botulinum* can be achieved by a combination of pH, acidity and aqueous salt content (both sodium chloride and emulsifying salts). Although not intended for ambient storage after opening, the cheese sauce could have achieved ambient safety if the water activity was reduced to 0.94 by increasing the aqueous salt content to 10% or greater or, indeed, by decreasing the pH to 4.6 or less. However, as the product was suitably ambient stable in relation to its intended purpose, i.e. prior to opening, and as it was not formulated to be ambient stable after opening, as is the case for many canned, low-acid foods, the only factors ensuring safety after opening would have been the avoidance of contamination and chilled storage.

The clearest message from this outbreak is that full consideration should be given to the information that must be supplied with the product to ensure that those who use such a product are made aware of the control they must exercise to ensure safety is maintained through to the final consumer. Importantly, the message about the product's instability after opening should have been made 'loud and clear' for everyone to understand. Of course, it is then essential that those using the product do so in accordance with the instructions on the product.

CANNED SALMON: UK

In Birmingham, UK, in 1978, four elderly patients, two females and two males, suffered symptoms of botulism within 18 hours of consuming afternoon tea together (Ball *et al.*, 1979). Initial symptoms of nausea, vomiting, dry mouth, dizziness and blurring of vision occurred with all individuals within nine to eleven hours of consumption of the implicated food. Within 14 hours, all four individuals had been admitted to hospital and rapidly developed paralysis, after which they were transferred to intensive care, only 18 hours after consumption of the food. Despite supportive treatment, two of the patients died (aged 64 and 66) after 17 and 23 days respectively. The other two patients were discharged from hospital after 75 days.

This botulism outbreak was caused by *C. botulinum* type E and toxin was detected in the serum of the patients. Sufficient toxin was present to result in the death of one out of two mice inoculated with 0.4 ml of a 1/64 dilution of the patients' serum.

The products consumed by all four patients at the afternoon tea included a tinned salmon salad, fruit and cream. The implicated food was the tinned salmon, remnants of which yielded spores of bacilli under microscopy and colonies of *C. botulinum* type E after culture (Table 2.8). *C. botulinum* was not detected in any of the other foodstuffs retrieved from the patients' home. The salmon was tested for the presence of botulinum toxin by injecting a saline suspension of fish residues from the implicated can into a pair of mice, both of which died within two-and-a-half hours. Mouse protection studies demonstrated that the toxin was that of *C. botulinum* type E (Ball *et al.*, 1979).

The tinned salmon was part of a batch of 14 000 cans manufactured in July 1977 in Alaska and imported to the UK from the USA. Although the product was purchased and consumed over one year after manufacture this is not unusual for canned ambient stable foods. Tins retrieved from the same batch and subsequently tested showed no signs of contamination and the incident appears to have been restricted to the single can of salmon. The cannery in Alaska is believed to have been operating to acceptable standards of heat processing and no evidence of under-processing of the product has been reported. Products of this nature would be given heat treatments in excess of 121°C for three minutes which would readily achieve a 12 log reduction in spores of *C. botulinum*. However, it is reported that operators involved in the gutting and preparation of the raw fish in the can processing factory may have been

Table 2.8 Outbreak overview: canned salmon

Product type:	Canned salmon
Year:	1978
Country:	UK
Levels:	Levels not reported
Cases:	4 (2 deaths)
Type:	*C. botulinum* type E

Possible reasons

(i) Presence of tear/weakness in the can seam and/or damage during subsequent processing, handling or transportation

(ii) Post heat process contamination during cooling, conveying, storage or sale

(iii) No evidence of growth of *C. botulinum*, e.g. blown can, due to escape of gas from 'hole' in can

Control options*

(i) Can integrity checks during manufacture using statistical process control techniques

(ii) Processing to avoid potential for can damage during cooling and subsequent transportation/storage/display

(iii) Post process can integrity checks

(iv) Elimination of sources of contamination to cooling cans

*Suggested controls are for guidance only and may not be appropriate for individual circumstances. It is recommended that proper hazard analysis is carried out for every process and product to identify where controls must be implemented to minimise the hazard from *C. botulinum*.

drying wet gloves and aprons on the warm cans which were cooling after processing (Stersky *et al.*, 1980).

Canned fish products are usually produced by first pre-cooking the whole gutted fish to mild pasteurisation temperatures of 60–70°C. The fish is allowed to temper and then it is portioned and placed into cans for further processing. The fish may be covered in water or a weak brine solution. The lid is applied and sealed to the can. During retorting at high temperatures, the closure sealant softens and, if seals are inadequate, can present an entry point for contaminants during subsequent stages of the process. Upon cooling, the can contents contract creating a vacuum and any defective seals or any holes in the seam or can could allow water or air to be drawn in along with microbial contaminants. It is apparent from subsequent investigations of the implicated salmon can that there was evidence of damage to the can which may have allowed ingress of contaminants. It was speculated that damage to the can in the area of the seam may have created a hole sufficient for post-process contaminants to gain entry, probably during cooling. The placing of clothing which may be heavily contaminated with microorganisms from raw fish processing onto cooling, processed cans is clearly a possible source of the organism. *C. botulinum* will certainly be present occasionally on raw fish and this practice could clearly introduce such contaminants into the final product. Such practices should never occur in canneries.

That the canned salmon did not appear overtly spoiled upon opening after a period of one year may have been due to the nature of the *C. botulinum* type together with the presence of a hole/tear. As the organism was non-proteolytic it would not be expected, in isolation, to cause proteolytic breakdown, which is often associated with a putrid odour. Non-proteolytic strains of *C. botulinum* are saccharolytic (Table 1.1) and would be expected to produce gas leading to swelling of the can. However, the presence of a small hole or tear at the seam may have allowed any gas produced to escape. That the product apparently showed no organoleptic signs of spoilage to the consumers is a significant factor in this outbreak.

The evidence from this outbreak indicates that it was caused by an isolated toxic can of salmon that was possibly contaminated during cooling or subsequent storage/handling, with spores of *C. botulinum* entering via a small hole or tear at the seam of the can. The processing of any canned product cannot reduce to zero the risk of incidents such as these occurring. That only one can out of a whole batch was contaminated points to an unfortunate sequence of circumstances that must have prevailed to allow this to occur.

It is rare for food poisoning to be caused by underprocessing in commercial canneries as the usually high levels of process control are generally sufficient to prevent such obvious occurrences. Post-process contamination is the main area where things go wrong in canning and it is difficult to completely eliminate the chance of post heat process bacterial contaminants entering a defective can.

There are three main areas where cans could be damaged in commercial canneries thus facilitating post heat process contamination. Firstly, it is possible that the can or its side seam (if a seamed can is used) could be defective at the point of construction. Secondly, the seal between the can and the end closure could be inadequately aligned or sealed. Finally, the seams and seal could be adequate but can damage may occur on conveyors, during cooling or even during subsequent transportation and retail handling. Defects or damage arising from any of these causes could allow microbial contaminants to enter the can contents after processing. It is also possible that deficiencies can be introduced into the can during can reforming which was a common, historical industry practice.

Cans arrive at a cannery in a number of different forms; usually as an intact unit with the bottom and side seam in place but it was common practice for can 'bodies' to be sent to manufacturers in a flat metal state as it increases transport efficiency; flat metal occupies much less space than formed cans. Prior to manufacture, the can is formed and then filled. A closure is applied, sealed and then the canned product is heat processed.

Can reforming must be controlled very precisely as it has the potential to introduce deficiencies into the can (or even tears) which either immediately, or after some transportation stress, allow entry of microbial contamination. Can and seam integrities are absolutely critical to the safety of these types of products.

Because the raw material may occasionally contain spores of *C. botulinum* and an ambient stable product is required, a heat process of F_o3 or more is applied to these products to provide an assurance of safety. Products like canned salmon contain few inhibitory factors to the growth of surviving pathogens whether they are spore formers or not.

Can integrity together with seal and seam checks should be conducted at frequent intervals in canneries to ensure that the defined tolerances of all critical measures of can, seal and seam integrity are complied with. Use of statistical process control techniques, to ensure that specifications remain under strict control, are often employed in the area of seam and seal checks. However, even with complete control of the machinery for the manufacture of cans and can closure devices, it is not possible to completely eliminate the chance of a defective unit occurring. However, the main aim, like the application of a heat process, is to reduce to an extremely low level the chances of a defect occurring. With inappropriate control of handling systems after the heating process however, it is possible to create significant problems leading to can leakage. This is usually attributable to can damage during conveying in or after the cooling process.

Cans are especially vulnerable to damage during cooling because the sealant is more fluid after being exposed to high temperatures. Collisions between cans as they are conveyed can create areas of stress that may lead to the seal being compromised and a pin hole developing. Such occurrences are far greater in cans which have apparent damage, particularly in the area of the seam and closure. Cans with overt signs of damage at this stage should be removed and discarded.

Post-process leakage can result in contaminants entering the can during the cooling stage due to the vacuum effect caused by the heat expanded product contracting during cooling and drawing in cooling water or air. Systems for cooling cans which maintain high pressure with air are often used to avoid distortion of cans or can seams during this stage.

Contaminants entering the can may be minimised by maintaining high standards of hygiene in the post heating processing plant. Use of non-potable water for cooling or for cleaning purposes can lead to heavily contaminated water entering defective cans during cooling. In addition, wet, poorly cleaned conveyor lines used to transport cans to packing areas can be a significant source of contamination which, again, could enter cans. Water used for cooling should be chlorinated. Chlorine has been shown to reduce numbers of *C. botulinum* spores, although this is highly dependent on the levels of contamination and the presence of other organic contaminants in the water that may inactivate the free chlorine. Ito and Seeger (1980) demonstrated that reductions of spores by some 99.99% could be achieved for all types of *C. botulinum* if chlorine exposure was maintained at 4.5 ppm for 3-12 minutes. Thus, ensuring that levels of chlorine are maintained at sufficiently high levels in cooling water and that such levels are not depleted by organic debris are critical components in the safety of canned products. Regular monitoring of free chlorine levels and frequent changes of cooling water are therefore also important considerations.

Under no circumstances should equipment or clothing from the raw side of the cannery come into contact with the cooling canned product and, although it was unclear as to whether this was the cause of the outbreak under discussion, the exposure of cooling cans to raw fish contaminants would clearly be regarded as poor practice.

This outbreak was clearly a 'one in a million' incident which is very difficult to rule out altogether. However, the significant advancements in can forming technology since this outbreak together with strict adherence to post-heating hygiene controls and finished product can integrity checks can greatly reduce the chances of a recurrence.

SALTED AND AIR-DRIED FISH: USA AND ISRAEL

Different eating habits throughout the world significantly affect the types of botulism outbreaks occurring in different countries. Most countries have their own traditional foods, the manufacture of which is historically based rather than based on modern food safety principles. Some of these traditional foods, manufactured in the home or in semi-commercial small-scale manufacturing units, can be the cause of botulism outbreaks. Native Alaskan foods, for example, which involve minimal processing of meat from aquatic mammals, have been the source of many *C. botulinum* type E outbreaks over the last century. Cold smoking, fermentation, drying or salting of fish, fish eggs or marine mammals such as whale or seal meat combine to make the incidence of type E botulism high in native Alaskans.

The vulnerability of traditional fish foods to the growth of *C. botulinum* was clearly demonstrated in an outbreak of type E botulism which occurred in 1987 due to the consumption of a traditional salted and air-dried fish product.

A total of eight cases were reported to have occurred of which six were identified in Israel and a further two cases in New York, USA. The affected individuals all suffered varying symptoms of botulism and one person died (Kotev *et al.*, 1987). The implicated food was a commercially produced, salt preserved, air dried, white fish known as Kapchunka or Ribyetz (Table 2.9). The implicated product and a serum sample from one of the patients were found to contain type E botulinum toxin.

Kapchunka is processed without evisceration and is undoubtedly susceptible to the growth of *C. botulinum* as it has been reportedly implicated in two previous outbreaks of botulism. In 1985, again in New York, two people died from type E botulism after consumption of Kapchunka (Bell *et al.*, 1985) and, in 1981, one case of botulism was reported in California, USA (Anon, 1981).

Kapchunka is a commercially produced product generally made by first dry-salting uneviscerated white fish and leaving it in a saturated brine solution under refrigerated conditions for periods of up to 28 days (Telzak *et al.*, 1990). During this period the salting elevates the aqueous salt content throughout the fish. This stage is critical to product safety as the next stage of the process involves rinsing salt off the fish and drying at ambient temperature for 3-7 days prior to sale and consumption of the flesh either raw or cooked.

Table 2.9   Outbreak overview: air-dried uneviscerated fish

Product type:	Salted, air-dried uneviscerated white fish
Year:	1987
Country:	Israel and USA
Levels:	Levels not reported
Cases:	8 (1 death)
Type:	*C. botulinum* type E

Possible reasons

(i)   *C. botulinum* spores in the fish gut microflora

(ii)   Germination of spores and growth in the gut due to anaerobic conditions and inadequate or slow salt diffusion into the fish gut during refrigerated brining

(iii)   Growth and toxin production in the fish gut due to inadequate salt concentration prior to air drying at ambient temperature

Control options*

(i)   Ensure adequate brine diffusion throughout the fish and fish gut to achieve > 3.5% aqueous salt during refrigerated brining

(ii)   Brining for shorter periods (< 10 days) or at lower temperatures (< 3°C)

(iii)   Brine injection to elevate aqueous salt in gut and flesh rapidly

(iv)   Ensure aqueous salt content 10% or greater during air drying at ambient temperature

*Suggested controls are for guidance only and may not be appropriate for individual circumstances. It is recommended that proper hazard analysis is carried out for every process and product to identify where controls must be implemented to minimise the hazard from *C. botulinum*.

The exact details of the process involved in the manufacture of the implicated product are not clear but it is reported that, although the salt concentration in the fish viscera was not measured separately, the aqueous salt content of the fish (final product) in the 1987 outbreak ranged from 18 to 24%. Levels of 10% or more would be expected to prevent the growth and toxigenesis of *C. botulinum* and it is therefore likely that the growth and toxin production of the organism occurred at one of the three key manufacturing stages of the product, i.e. distribution of the uneviscerated fish to the processing facility, in the viscera of the fish during salting or in the viscera of the fish during air drying. It is not possible to rule out any of these stages as contributors to the problem that occurred.

Spores of *C. botulinum,* particularly type E, are frequently reported to be present in marine sediments and this type is the most likely to be present on/in fish and marine animals (Tables 1.6 and 1.7). Its presence in the gut microflora of fish should therefore be anticipated as a potential hazard for these types of products.

Type E *C. botulinum,* responsible for these outbreaks, is capable of psychrotrophic growth. The temperature required to prevent growth in foods (<3.3°C) is unlikely to have been maintained and growth could easily be envisaged both during transport of the raw fish and during the brining process. Conditions in the gut of the fish would undoubtedly be conducive to the growth of anaerobic bacteria like *C. botulinum* and salt penetration is likely to be slowest to this region. Therefore, it is certainly theoretically possible that growth and toxin production could occur, particularly given the extended salting period. It is also possible that toxin production may have occurred in the gut and diffused into the flesh (Telzak *et al.*, 1990).

To prevent growth of *C. botulinum* and toxin production at the brining stage, it would have been necessary to achieve either a temperature below the threshold for growth of non-proteolytic strains, widely accepted as 3.3°C in foods (it has been shown to grow in broth at 3.0°C, see Chapter 3), or to rapidly achieve an aqueous salt concentration throughout the fish in excess of 3.5% to delay or prevent growth during the 28-day brining process. Although the upper salt concentration allowing growth of non-proteolytic strains of *C. botulinum* is generally believed to be 5% (Table 3.6), at concentrations of 3.5% aqueous salt, growth would not occur to hazardous levels over 28 days providing the temperature was maintained below 5°C.

Although there was the potential for psychrotrophic growth and toxin production by *C. botulinum* type E during salting, it is equally possible that growth and toxin production could occur during the drying stage carried out at ambient temperatures. Under such conditions, levels of 10% aqueous salt would be necessary to prevent growth of proteolytic strains (types A, B and F) and over 5% for non-proteolytic strains (types B, E and F). Despite the measured salt concentration of > 18% in the final product in the 1987 outbreak, high levels may not have been attained in the fish gut either at all or rapidly enough to prevent growth and toxigenesis by *C. botulinum*. Even at moderate ambient temperatures of 20°C, *C. botulinum* type E can readily increase to toxin forming levels within 3–7 days if the salt concentrations are not inhibitory. Of course, the potential for growth of proteolytic strains of *C. botulinum* is even greater at ambient temperatures as their tolerance of salt is significantly greater than that of the non-proteolytic strains.

In lessons to be learnt from this outbreak it is clear that simple rules need to be followed to prevent a future recurrence of these types of incidents. Hazard analysis techniques and appropriate process controls can be

applied in the production of traditional foods without significant altera-
tion of the traditional process involved in their manufacture. What is
required is the clear conveyance of such messages to those making these
foods.

Firstly, ensuring that the raw fish is kept for as short a period of time as
possible under good refrigeration temperatures prior to the salting stage
would prevent the growth of proteolytic strains of *C. botulinum* and
reduce the rate of growth of non-proteolytic strains. Salting at temperatures below 10°C would inhibit the growth of proteolytic strains while
salting at < 3°C would prevent the growth of non-proteolytic strains.
Alternatively, reducing the time of salting from 28 days to < 10 days would
significantly reduce the likelihood of growth of non-proteolytic strains and
prevent toxin formation during brining. However, it should also be
remembered that lowering the temperature or reducing the salting time
may interfere with salt penetration and equilibration and therefore make
the subsequent stage at elevated temperature more hazardous. Therefore,
it is important to assess the minimum times, temperatures and brine
concentration required to achieve sufficiently high salt concentrations
throughout the fish capable of preventing growth of *C. botulinum* during
air drying.

An alternative course of action could be to introduce salt directly into the
flesh and intestine by brine injection which would rapidly increase the
speed of salt elevation in the areas vulnerable to growth of *C. botulinum*.

Finally, prior to elevation of temperature during air drying, the salt content
in the flesh and gut should be monitored to ensure that a minimum of 10%
in the aqueous phase is achieved throughout the fish to prevent the
growth of proteolytic strains of *C. botulinum*. Although 5% aqueous salt
would have been sufficient to prevent growth of the non-proteolytic
strains, it is clear that once the temperature is elevated to within the
growth range of the proteolytic types, and as these are more tolerant of
high salt concentrations, they represent the hazard most in need of
control at this stage.

Of course, the extent to which any of these factors can be applied and
controlled under traditional 'home-made' conditions is debatable. Nevertheless, these factors are clearly within the capability of larger scale
manufacturers who can turn a traditional product with marginal food
safety characteristics into one that is of more assured food safety
characteristics.

MEAT PIE: USA

An unusual case of botulism occurred in the US in 1982. The incident, although attributable to a commercially produced food, was caused by inadequate controls in the home. One person was affected and she developed typical symptoms of botulism one day after consumption of the food. The patient was hospitalised and required intensive care treatment. The organism responsible was *C. botulinum* type A, the toxin of which was detected in the serum of the patient. The implicated product was a beef pot pie (Table 2.10) and *C. botulinum* type A toxin was detected in an uneaten portion of the pie retrieved from the patient's home (Anon, 1983).

The pie was manufactured as a processed frozen meat pie; such products would normally be made by mixing the ingredients and then cooking the mix in a large vessel at temperatures up to 100°C until the meat pieces are tender. The filling is then deposited into pastry cases either hot or after cooling quickly and storing chilled until required. The cooked filling (hot or cold) is filled into raw pastry cases and the pie may then be baked to

Table 2.10   Incident overview: meat pie

Product type:	Manufactured frozen meat pie cooked in the home	
Year:	1982	
Country:	USA	
Levels:	Levels not reported	
Cases:	1 (no deaths)	
Type:	*C. botulinum* type A	

Possible reasons
(i)   *C. botulinum* spores not destroyed by manufacturing processes
(ii) Surviving *C. botulinum* spores in the meat pie not destroyed by oven baking in the home
(iii) Germination and growth of spores in the pie during unrefrigerated storage in the home after baking
(iv) No recooking of the pie after temperature abuse and prior to consumption

Control options*
(i)   Consumer education and awareness
(ii) Clear instructions indicating the product should be consumed immediately after cooking

*Suggested controls are for guidance only and may not be appropriate for individual circumstances. It is recommended that proper hazard analysis is carried out for every process and product to identify where controls must be implemented to minimise the hazard from *C. botulinum*.

produce a ready-to-eat or heat product, or it may be frozen directly after filling for full home baking.

There were no indications that the pie had not been manufactured safely and pre-formed toxin was not considered as a likely factor in this incident. However, the processes employed in the production of meat pies are clearly insufficient to destroy spores of *C. botulinum*. As reported (Anon, 1983), the frozen pie was taken by the son of the patient two-and-a-half days before she consumed it and cooked in an oven for 40–45 minutes. As he was about to serve the pie, his father returned home with some 'take-away' hamburgers and so the pie was placed on an unrefrigerated shelf. Two and a half days later, on returning home, the son noticed that his mother had consumed some of the pie without reheating it.

Clearly, this incident was caused by an unfortunate accident which led to extensive temperature abuse of the product. Cooking processes employed by the manufacturer of the pie would have been designed to achieve meat tenderisation and would also destroy vegetative pathogens. However, pies of this nature will usually be made with pastry that is raw or only partly cooked and the expectation is that the consumer will expose the product to a full bake. In fact, it seems that this is indeed what the patient's son did, but as the manufacturing process and the subsequent bake were incapable of destroying spores of *C. botulinum* type A, the subsequent temperature abuse would have been sufficient to allow extensive growth and toxin production by surviving spore-forming bacteria.

Following manufacture, the pie was frozen and then distributed and sold in a frozen state. No organism growth will occur in stable freezing conditions and freezing acts as an effective preservation method for otherwise perishable foods.

Although the pH and water activity of the implicated pie are not reported, pies of this nature usually have a very high water activity and neutral pH. The composition may include some salt, bringing the water activity to perhaps 0.98, but as no other effective preservation mechanisms would be present in the defrosted cooked pie, the only true potential controlling factors were the temperature and time after consumer cooking. As the pie remained at warm or ambient, but not hot temperatures, from initial cooking to the time, two-and-a-half days later, when the patient ate some of it, it is likely that for much of the time the temperature of the pie was within the optimum growth range for *C. botulinum* of 20–40°C.

Preventing botulism incidents of this nature is very difficult indeed. Clearly, it is not possible to manufacture foods of this nature without the potential for spores of *C. botulinum* to be present as they are likely to be present in the raw materials used. These would then survive the cooking process. Meat, herbs, spices and flour are some of the major ingredients in these types of product and all have the clear potential, from time to time, to carry with them spores of *C. botulinum*. The meat pie, once made, could be considered to be an optimum growth medium for *C. botulinum* and, therefore, the manufacturer and retailer of the product are as reliant on the customer to cook the product properly and to avoid subsequent temperature abuse as they are on their own systems of control in the factory, distribution and retail chain.

It is, of course, common sense that food should not be left under warm conditions for extended periods prior to eating. However, it may not be clear to the general public how quickly food poisoning bacteria can grow in some foods and therefore how important time can be in relation to their 'leftovers'. In such situations it is important for everyone to receive some education in food hygiene and food safety so that the basic messages about keeping hot foods hot or cooling them down quickly and keeping them in the refrigerator are clearly understood.

It is not important for the general public to understand the detail relating to the specific hazard of *C. botulinum* but it is essential that they recognise that some food poisoning bacteria are not destroyed by cooking alone and can grow in warm foods very quickly. The responsibility for such education must be via government health and education departments. This should be supplemented by manufacturers and retailers of food who can complement the basic education by providing appropriate reminders to their customers on product packaging or in the form of food safety leaflets regarding the important aspects of food hygiene and safety.

Unfortunately, incidents of this nature are not confined to botulism and many outbreaks of salmonellosis have occurred over the years due to temperature abuse of foods in the home because of poor understanding of the inherent hazards. It is not possible for the food industry alone to completely control food safety; there is a clear responsibility for all of those in the food chain, from producer and manufacturer to retailer and consumer, to understand their particular role in the process to ensure safety throughout the chain. This may be an unrealistic goal in some poorly developed countries in the world today but in developed countries, where good education systems exist, such messages could clearly contribute to improve public health.

3

FACTORS AFFECTING GROWTH AND SURVIVAL OF *CLOSTRIDIUM BOTULINUM*

GENERAL

Despite the widespread distribution of *C. botulinum* spores in the environment and consequently in and on raw foods, it has long been understood that, with the exception of infant botulism, foodborne botulism occurs following multiplication of the organism and toxin production in the food prior to consumption and not as a result of germination of spores and growth within the alimentary tract. Outbreaks of botulism are most commonly associated with foods which have been insufficiently heat processed to destroy spores and/or where conditions in the food and its subsequent storage conditions are favourable to the growth of the organism.

The spores of *C. botulinum* must be expected as contaminants of raw foods. As it is impossible for raw agricultural and aquatic produce to be free from *C. botulinum* spores at source, it is important to establish animal husbandry and crop agricultural regimes aimed at minimising the soil and faecal contamination load on raw material foods reaching food processing plants. Thereafter, raw material handling and processing procedures need to be structured and operated to prevent the germination of any *C. botulinum* spores present and minimise any multiplication of the organism.

In addition to attention to the detail of normal cleaning and hygiene procedures, the treatment processes and formulation of food products are important for controlling any residual *C. botulinum* organisms and preventing their potential to cause harm to consumers.

It should be remembered that apparent anaerobic conditions are not a pre-requisite for the germination and growth of *C. botulinum* spores and the organism may grow in air-packed foods as well as in modified air or vacuum-packed foods (Hauschild, 1989). This is because the prevailing oxidation–reduction potential or redox potential (Eh) in the food constitutes a potential growth controlling factor whether the pack atmosphere is apparently aerobic or anaerobic.

In addition to Eh, which can fluctuate during shelf life, foods consist of a multitude of micro-environments and it is such environments, even in aerobically packed food, which can provide conditions in which *C. botulinum* may grow. Sugiyama (1982) refers to work in which growth of *C. botulinum* type A and toxin production was observed in inoculated fresh over-wrap packed mushrooms held at 20°C for six days. This occurred because the respiratory processes of the mushrooms reduced the oxygen in the packs to levels at which the inoculated spores could germinate, grow and produce toxin. Perforations in the over-wrap material allowed oxygen equilibration at high enough levels to prevent toxin production (Sugiyama and Rutledge, 1978).

Although a strict anaerobe which grows optimally at an Eh of $-350\,mV$, growth of *C. botulinum* may be initiated in the Eh range $+30\,mV$ to $+250\,mV$ (Kim and Foegeding, 1993) subject to type and other physico-chemical conditions prevailing. These conditions can be significantly influenced by the growth of competitor organisms. *C. botulinum* spore germination can occur at higher Eh values than those at which growth can occur and the metabolic activity during germination can reduce the Eh (Kim and Foegeding, 1993). During the growth of competitor organisms, oxygen is depleted; reducing compounds such as thiols are produced simultaneously and this activity can lead to a reduction in Eh to levels at which *C. botulinum* may grow. Eh conditions conducive to growth of *C. botulinum* may also occur just millimetres below the surface of fish or meat muscle (Mossel *et al.*, 1995). It is little surprise therefore that delicatessen food products in packs only approximately 3 cm deep have been shown to have environments with Eh ranging from $-198\,mV$ to $-23\,mV$, levels which may well allow germination and growth of spores of proteolytic and non-proteolytic *C. botulinum* (Snyder, 1996).

In theory therefore, most foods may have micro-environmental conditions suitable for the growth of *C. botulinum*. In practice, however, the combined conditions within foods, processes applied, storage conditions and shelf life have served to prevent frequent outbreaks of botulism. However, such information does underline the need to ensure that technical personnel in the food industry have a thorough understanding of the factors combining to make a particular food product safe from the hazard of *C. botulinum*.

Within food production processes, a variety of physico-chemical factors used either singly or in combination can be effective in controlling the survival and growth of *C. botulinum* during processing and also in the finished food products.

TEMPERATURE

Heating processes are among the most important of the treatments applied to foods and the proper control of heating processes is crucially important in helping prevent outbreaks of botulism. Vegetative cells of *C. botulinum* are as heat sensitive as the vegetative cells of most other bacteria and are readily destroyed by normal pasteurisation temperatures (International Commission on Microbiological Specifications for Foods, 1996). It is the heat resistance of their spores which is of primary concern for food safety. The heat resistance of *C. botulinum* spores was one of the first of the physico-chemical characteristics of the organism to be investigated in detail with a view to applying the results in food production processes.

Esty and Meyer (1922) studied the heat resistance of 109 strains of *B. botulinus* (sic) obtained from soils and vegetables in addition to foods implicated in outbreaks of human and chicken botulism. They found a wide variation in heat resistance between strains and discovered that some strains sporulate irregularly and poorly. Reasons for these observations were investigated and, although the medium and method of spore production were found to influence heat resistance, other unknown factors were believed to exist. Despite the problems encountered, Esty and Meyer provided much of the key data upon which food heat processes are based today. Tables 3.1 and 3.2 show the maximum heat resistance of *B. botulinus* (*C. botulinum*) spores artificially produced under ideal growth conditions and the results of some experiments to determine the relationship between spore concentration and heat resistance.

Esty and Meyer also studied the effect of varying concentrations of sodium and potassium chlorides (NaCl and KCl) in a veal infusion medium on the destruction time (*DT*) of the spores of *C. botulinum*. Destruction times at 100°C were longer in the presence of 0.5% and 1.0% NaCl than 2.0% and

Table 3.1 Maximum moist heat resistance of *B. botulinus (C. botulinum)* spores artificially produced under optimum growth conditions: spores heated in a phosphate solution at an approximate pH 7.0, adapted from Esty and Meyer (1922)

Temperature (°C)	Time (minutes)
100	330
105	100
110	33
115	10
120	4

3.0% NaCl with no further reduction in destruction time occurring until the concentration reached 8.0% (2.0% and 4.0% NaCl, $DT = 35$ minutes, 8.0%, $DT = 30$ minutes and 20.0%, $DT = 12$ minutes), although it should be noted that the recorded pH of the infusions decreased from 5.94 to 5.06 over the range 0.5–20.0% NaCl and this may have had a synergistic effect in reducing destruction times. The pH of food also has an important influence on the efficacy of a heat process. In some of their early studies, Esty and Meyer reported the maximum heat resistance of favourably

Table 3.2 Relationship between concentration of spores and heat resistance, adapted from Esty and Meyer (1922)

Spore strain	Spores/cc	Resistance at 105°C	
		survival (min)	destruction (min)
90	9×10^8	44	48
	9×10^6	34	36
	90 000	18	20
	900	12	14
	9	—	2
Spore strain	**Spores/cc**	**Resistance at 100°C**	
		survival (min)	destruction (min)
97	72×10^9	230	240
	1.64×10^9	120	125
	32.8×10^6	105	110
	650 000	80	85
	16 400	45	50
	328	35	40

Destruction time (*DT*): The time necessary to destroy a known suspension of spores under specified conditions.

treated spores at pH 7.0 to be 330 minutes at 100°C (Table 3.1) and also that the heat resistance of spores of *C. botulinum* tends to decrease with reducing pH (Table 3.3), a marked decrease occurring at pH values < 4.5. More recent work demonstrated a reduction in the $D_{107.2}$ value for spores of one strain of *C. botulinum* to be 4.7 minutes at pH 7.0, 3.0 minutes at pH 6.0 and 2.3 minutes at pH 5.0 (Hutton *et al.*, 1991).

Table 3.3 Destruction times of *C. botulinum* spores in acidified spinach juice; adapted from Esty and Meyer (1922)

Acidified with hydrochloric acid		Acidified with citric acid	
pH	Destruction time at 100°C (min)	pH	Destruction time at 100°C (min)
Control 5.05	50	Control 5.26	65
4.80	50	4.92	50
4.50	45	4.77	45
4.38	40	4.69	40
4.31	26	4.54	35
4.16	22	4.5	35
4.11	20	4.37	30
3.98	15	4.34	25
3.81	15	4.31	25
3.70	15		

During their work on resistance of spores in different food juices, Esty and Meyer (1922) noted that spores exhibited very different heat resistance to that observed in pH modified artificial media, sometimes with very little difference being observed in heat resistance in food juices of markedly different pH. For example, approximately the same resistance was found in juice from ripe olives (pH 7.93), corn (pH 6.35) and spinach (pH 5.05). Their advice, given in 1922, to carefully consider the implications of these findings when determining the processing conditions applicable in food canning, remains sound advice today.

The clear potential for the spores of strains of *C. botulinum* to survive heat processes is clear and, as a result of such work, it is now widely accepted that for low- and medium-acid, ambient stored foods such as vegetables and uncured meats which do not inhibit the growth of the organism, standard canning industry heat processes should achieve a reduction in the population of *C. botulinum* spores by a factor of 10^{12}; known as a 12*D* process or the 'botulinum cook'. The most resistant spores of *C. botulinum* have been found to have a *D* value in neutral phosphate buffer at 121.1°C (250°F) of 0.21 minute, so a 12*D* process for these organisms at this temperature would be achieved in 2.52 minutes; 3 minutes in commercial practice (assuming instantaneous heating and cooling) (Hersom and Hulland, 1980). *D* values do vary depending on the medium (food) to which the heat process is applied (Table 3.4) and, for food processes, structured challenge studies may need to be carried out to establish the *D* value and *z* value of relevant strains of *C. botulinum* in the particular food of concern and thus ensure a safe canning heat process is applied.

It is particularly important to review the efficacy of heat processes when any changes are made to product formulation, e.g. ingredients and their physico-chemical characteristics including water activity (solute types and levels) (O'Mahony *et al.*, 1990), pH (type of acid), fat content, particle size or consistency, can/container size, pre-processing conditions, and processing equipment used (including equipment modifications).

Because of the significant development of the market in cooked-chilled foods, vacuum and modified atmosphere packaged-chilled foods, particularly over the past two decades, and the commercial pressure to obtain shelf lives which are as long as possible for these products, there has been an increased focus of attention on the implications of psychrotrophic, non-proteolytic *C. botulinum* in such foods.

The spores of the non-proteolytic types of *C. botulinum* are less heat resistant than the proteolytic types (Table 3.4); however, resistance is still high enough to allow survival of many heat processes in which the maximum temperature reached in foods is < 100°C. From work commissioned by the UK Ministry of Agriculture, Fisheries and Food and carried out at Campden Food and Drink Research Association (now Campden and Chorleywood Food Research Association, CCFRA), a *D* value at 90°C for a psychrotrophic *C. botulinum* type B spore population in cod was found to be 1.1 minutes (*z* value = 9.0). Based on this work, the temperature/time combination to achieve the required six decimal reductions (6*D*) of this spore population was calculated as 90°C for 7 minutes (Advisory Committee on the Microbiological Safety of Food, 1992).

Because *D* values vary according to strain of organism, substrate in which the spores are heated, methods and recovery media used, a safety margin was built into the heat processes required to achieve a six decimal reduction in a population of psychrotrophic *C. botulinum* spores using a temperature of 90°C; 10 minutes is the specified time period at this temperature. From this 6*D* process, which includes a safety margin, equivalent temperature/time processes have been calculated (Table 3.5).

Peck (1999) reviewed the implications of the natural presence of lyso-zyme in raw foods. Heat treatments can cause sub-lethal injury to spores of non-proteolytic *C. botulinum* by inactivating the spore germination sys-tem and some heat damaged spores are permeable to lysozyme which can diffuse through the spore coat and induce germination by hydrolysing the peptidoglycan in the spore cortex. Fernandez and Peck (1999) found that few or no spores were recovered in the absence of lysozyme following heat treatments of 85°C to 95°C but, in the presence of lysozyme, sub-stantial numbers of spores germinated and grew. Growth was also more rapid over a wider range of conditions when hen egg white lysozyme was present than in the absence of lysozyme. Thus, the natural presence of lysozyme in many foods such as eggs and fish may significantly increase the heat process conditions required to inactivate the spores of any non-proteolytic *C. botulinum* that may be present. Therefore, for such foods, a process of 90°C for 10 minutes may not achieve the generally required 6 log reduction in spores of non-proteolytic *C. botulinum*. However, where such heat treatments are followed by controlled refrigerated storage conditions, <8°C, germination and growth from non-proteolytic *C. botulinum* spores is prevented or significantly delayed (Fernandez and Peck, 1999).

Botulinum toxins are destroyed by heat, e.g. temperatures of 80°C for 10 minutes or boiling for a few minutes, but, to be effective, the heating process must be thorough, achieving the required temperature through-out the food. Because cooking and re-heating processes are often not well controlled, cooking carried out by the consumer must not be relied upon to destroy botulinum toxin. In addition, a heat process used in the man-ufacture of the product must not be used as a means to 'clean up' unsafe raw materials.

As with canning processes, where a heat process is relied upon for the safety of a long-life, chilled product in respect of psychrotrophic *C. botulinum*, process efficiency reviews may need to be undertaken when changes are made to product formulation, e.g. increase in fat content or portion size, packaging or associated relevant processes.

Psychrotrophic *C. botulinum* strains are capable of growth at refrigera-tion temperatures (Table 3.6). The key reference to the now widely accepted minimum growth temperature of 38°F (3.3°C) is that of Schmidt *et al.* (1961) who, in their experiments, inoculated four different strains of *Clostridium botulinum* type E into tubes containing a heat sterilised beef stew (100 000–400 000 spores per g). Sets of tubes were incubated at 34°F, 36°F or 38°F and examined periodically for evidence of growth

Table 3.4 Some decimal reduction times (*D* value) measured for the spores of *C. botulinum* types A, B and E in experiments with various foods, adapted from International Commission on Microbiological Specifications for Foods (1996)

Temperature (°C)	Medium	*C. botulinum* type *D* value (minutes)/*z* value (°C)		
		A (proteolytic)	B (proteolytic)	E (non-proteolytic)
110.0	Pea purée	1.98/8.3	2.14–12.42/8.3	
115.6		0.44/8.3		
121.1		0.089/8.3		
100.0	Canned spinach, pH 5.39	17.26/10.0	22.13/8.6	
110.0		1.74/10.0	1.54/8.6	
100.0	Canned asparagus, pH 5.42	11.97/7.9	17.76/7.9	
110.0		0.61/7.9	1.06/7.9	
115.0		0.14/7.9		
121.0	Mackerel in water	0.15/10.6		
	Mackerel in oil	0.41/12.7		
110.0	Tomato juice, pH 4.2	1.50–1.59/9.43		
115.6		0.38–0.4/9.43		
110.0	Canned corn		2.15/9.6	
115.0			0.88/9.6	
120.0			0.24/9.6	
110.0	Mushroom purée, pH 6.3–6.5		0.49–0.99/—	
115.6			0.12–0.39/—	

Table 3.4 Continued

Temperature (°C)	Medium	C. botulinum type D value (minutes)/z value (°C)		
		A (proteolytic)	B (proteolytic)	E (non-proteolytic)
60	Oyster homogenate			776/7.5
70				72/7.5
80				0.78/7.5
73.9	Menhaden surimi			8.66/9.78
76.7				3.49/9.78
79.4				2.15/9.78
82.2				1.22/9.78
73.8	Crabmeat			6.2–10.8/10.5–11
76.6				1.5–2.8/10.5–11
79.4				0.7–1.3/10.5–11
82.2				0.4–0.6/10.5–11

Table 3.5 Equivalent time/temperature combinations for a
6*D* reduction in spores of psychrotrophic *C. botulinum*,
adapted from Advisory Committee on the Microbiological
Safety of Food (1992)

Temperature (°C)	Time (minutes)
70	1675
71	1290
72	1000
73	773
74	600
75	464
76	359
77	278
78	215
79	167
80	129
81	100
82	77
83	60
84	46
85	36
86	28
87	22
88	17
89	13
90	10

Calculated using z value = 9°C, reference temperature = 80°C.

using visible gas production as an indicator; tubes were also assayed for
type E toxin production.

No growth or toxin production was found with any of the strains tested
after incubation for 104 days at either 34°F or 36°F. Two of the four strains
tested were found to grow and produce toxin after 31 days' incubation at
38°F (3.3°C). The results from these experiments suggested that the
critical temperature below which growth and toxin production from
psychrotrophic strains of *C. botulinum* will not occur is 38°F.

Although growth of the organism at temperatures close to its minimum
38°F (3.3°C) is only slow, recent studies have demonstrated that psychrotrophic strains of *C. botulinum* can grow and produce toxin in some
foods within 7 days at 15°C and 12 days at 8°C (Table 3.7). In studies using
artificial growth media, although no growth was detected at 2.1°C,
growth and toxin production by non-proteolytic strains of *C. botulinum*

Table 3.6 Some physico-chemical characteristics of *C. botulinum*, adapted from Hauschild (1989)

Characteristic*	Group I	Group II
Minimum temperature for growth (°C)	10	3.3
Maximum temperature for growth (°C)	48	45
Inhibitory pH	4.6	5.0
Inhibitory sodium chloride (salt) concentration (%)	10 (water activity 0.94)	5 (water activity 0.97)

*Value when other conditions are optimal for growth.

were detected in 5-6 weeks at 3°C, growth occurring more frequently from spores of type F than types B or E (Graham *et al.*, 1997).

Chilled foods have been defined as 'perishable foods which, to extend the time during which they remain wholesome, are kept within controlled and specified ranges of temperature above their freezing points and normally below 8°C' (Anon, 1991a). Food production process, storage, distribution and retail display refrigeration temperatures are usually in the range 4-7°C and domestic refrigeration temperatures in the consumer's home have been found to exceed 10°C; all these temperature ranges include temperatures well above the minimum temperature for growth of psychrotrophic strains of *C. botulinum*.

It is for these reasons that advice has been given to the food industry to restrict the shelf life of prepared chilled foods that have no other psychrotrophic *C. botulinum* growth inhibiting factors, e.g. low pH, to a maximum of 10 days (Advisory Committee on the Microbiological Safety of Food, 1992).

Betts (1996) refers to chill temperatures as ≤8°C but, within this, two temperature ranges are discussed in respect of control of *C. botulinum* in vacuum-packed or modified atmosphere packaged foods. Growth of the organism will be prevented if these types of products are stored at 3°C or less throughout their shelf life; for such products stored at >3°C, other factors should be considered. At temperatures >3°C but ≤8°C, shelf life should be restricted to 10 days or less, in which case, no additional factors are necessary to control *C. botulinum* unless the hazard analysis process

Table 3.7 Time to toxin production of some types of psychrotrophic *C. botulinum* in different foods when incubated at chill temperatures, adapted from Betts and Gaze (1995)

C. botulinum type	Food	Temperature (°C)	Time to toxin production (days)
B	Cod (sous-vide)	5	70
		8	28
	Chicken (sous-vide)	8	42
	Cod (sous-vide)	15	7
	Chicken (sous-vide)	15	7
E	Cod (sous-vide)	5	42
	Chopped meat with 0.5% sodium chloride	6	14
	Cod (sous-vide)	8	21
	Chicken (sous-vide)	8	28
	Crab meat	10	8
	Potatoes in vacuum pack	10	9
	Crab meat	12	14
	Fish in vacuum pack	12	12
	Cod (sous-vide)	15	7
	Chicken (sous-vide)	15	7
B & E	Liver sausage with 83 ppm sodium nitrite	8	28
	Salmon fillets in modified atmosphere pack	8	12
	Tagliatelle with chicken or pork	8	28
	Liver sausage with 83 ppm sodium nitrite	10	18
	Salmon fillets in modified atmosphere pack	12	3

indicates otherwise. Those products in these categories assigned a shelf life of more than 10 days at chill temperatures ($>3°$C but $\leq 8°$C) will require additional controlling factors, e.g. reduced water activity, reduced pH, increased salt level.

The development of 'sous-vide' products in which raw and/or cooked foods are vacuum-packed, heat-treated using pasteurising temperatures, then given an extended refrigerated shelf life, has facilitated the commercial availability of a range of high organoleptic-quality, preservative-free, prepared meals. However, the anaerobic nature of these product packs, absence of vegetative competitive microflora due to in-pack pasteurisation, possible survival of *C. botulinum* spores, minimal (if any) preservative systems (other than chill temperature) and extended refrigerated shelf life combine to enhance the potential for psychrotrophic *C. botulinum* (if present) to grow and produce toxin in the food. As these organisms are non-proteolytic, obvious signs of spoilage may be absent when the food is toxic. This characteristic increases the hazardous nature of these types of *C. botulinum*.

Different types of *C. botulinum* have been shown to predominate in different geographical areas, environments and foods (Tables 1.4, 1.5 and 1.7). Therefore, for shelf-stable, low-acid canned or bottled foods and extended-life (> 10 days) chilled foods it is important to ensure that a thorough hazard and critical control point analysis of the full production process of the food takes account of the potential hazard from *C. botulinum* in general, and psychrotrophic strains in particular, in respect of extended shelf life chilled products.

pH, WATER ACTIVITY AND OTHER FACTORS

Table 3.6 indicates the generally accepted inhibitory pH values for strains of *C. botulinum* in Groups I and II. All types and strains of *C. botulinum* can grow and produce toxin to about pH 5.2 (all other conditions being optimal) but, under experimental conditions in homogenous protein-rich media, some strains (inoculated together with spores of *Bacillus* spp.) have been shown to grow and produce toxin at a pH as low as 4.0 with an inorganic acidulant and pH 4.4 with an organic acidulant (Raatjes and Smelt, 1979; Smelt *et al.*, 1982); such observations are thought to be due to micro-environments of higher pH formed by precipitated proteins. At pH 4.4, acetic acid was found to be more effective than lactic acid which was more effective than citric acid and, in turn, was more effective than hydrochloric acid in inhibiting the growth and toxin production by *C. botulinum*. Raatjes and Smelt did not find any growth of *C. botulinum* in a wide range of real foods with pH values less than 4.6 which remains the commonly accepted equilibrium pH value inhibiting growth of *C. botulinum* and toxin production and is used as the dividing pH between acid (pH ≤ 4.6) and low acid (pH > 4.6) foods (Odlaug and Pflug, 1978).

Outbreaks of botulism have occurred involving acid foods. Odlaug and Pflug (1978) summarised data from 35 outbreaks of botulism which occurred between 1899 and 1975 in the USA and attributed to acid foods; 34 of the 35 involved home-processed canned foods including pears, apricots, tomatoes, tomato juice, and pickles. The product pH was reported in only three of these outbreaks; pears, pH 3.86, tomatoes, pH 4.0 and tomato juice, pH 4.2. The single commercially produced acid product which was involved in an outbreak of botulism in 1915 was a tomato catsup.

Although acid canned foods are not sterile, the heat processes applied are designed to be sufficient to destroy all vegetative forms of bacteria, yeasts and moulds, some mould spores and some bacterial spores. The acid and anaerobic nature of the environment within a correctly prepared and treated product pack assures safety. However, process failures such as incorrect process design, incorrect heat process delivered or post-process contamination (including ingress of air) of acid canned foods can allow the survival of organisms present on the raw food which would normally be destroyed during processing or can introduce organisms during handling (Meyer and Gunnison, 1929; Bow *et al.*, 1974). The presence and growth of these contaminating organisms are believed to assist the germination of *C. botulinum* spores and their subsequent growth and toxin production.

Yeasts and lactic acid bacteria were found in home-canned spoiled Bartlett pears which caused two deaths due to botulinum toxin type A (Meyer and Gunnison, 1929). The recorded pH of the syrup was 3.86. It has been shown that different organisms, e.g. moulds and *Bacillus* spp., can grow in acid foods and raise the pH to a level above pH 4.6. Experimentally produced mould mycelial mats in tomato juice in non-hermetic containers were found to create a pH gradient; a higher, near neutral pH close to the mat and lower, acid pH further away from the mat. *C. botulinum* growth occurred near the mat but less so further away, although toxin was found throughout the product. In a hermetic unit, mould growth was restricted but *C. botulinum* growth and toxin production were still demonstrated and it was suggested that this took place close to or within the mycelial mat (Odlaug and Pflug, 1978). In real food situations, because pH gradients may be easily destroyed during physical movement of the container or even by gas bubbling through the spoiled product, the pH measured may not reflect that in the immediate vicinity of any *C. botulinum* growth which occurred. Such a possibility should be considered when evaluating data arising from investigations.

The potential for process survivors or post-process contaminating organisms to grow in an acid product and alter the conditions sufficiently, even in a localised area such as at the surface, to favour the growth of *C. botulinum*, if present, must be recognised and accounted for in a hazard analysis.

For food products without the contributory factor of a growth-inhibiting pH, some other *C. botulinum* controlling process or factor may be required. One of the most common alternative approaches to inhibiting the growth of *C. botulinum* is by the reduction of water activity by either drying, or the use of humectants, e.g. salt or sugar. The water activity required to inhibit the growth of proteolytic and non-proteolytic types of *C. botulinum* is shown in Table 3.6; however, this can be affected by the nature of the humectant (solute) used. The concentration of salt and associated water activity inhibitory to the growth of *C. botulinum* is also shown in Table 3.6. However, in relation to psychrotrophic *C. botulinum*, the UK Advisory Committee on the Microbiological Safety of Food (1992) recommended that in addition to chill temperatures which should be maintained throughout the chill chain, a minimum salt level of 3.5% in the aqueous phase (Table 3.8) throughout the food and throughout all components of complex foods should be used to prevent growth and toxin production in prepared chilled foods with an assigned shelf life of more than 10 days. It should be remembered however, that in otherwise optimal growth conditions, the minimum inhibitory salt concentration is 5%.

The inhibitory effect of salt is due mainly to the reduction in water activity caused; the higher the percentage concentration of salt in the aqueous phase, the lower the water activity, the greater the inhibitory effect on growth: 1% NaCl reduces water activity by about 0.006 units (Hauschild, 1989).

Table 3.8 Minimum salt level required for different water contents of food to obtain a minimum aqueous salt concentration of 3.5%

Salt level (% w/w)	Water content (% w/w)
1.82	50
2.00	55
2.18	60
2.36	65
2.54	70
2.73	75
2.91	80
3.09	85
3.27	90

$$\% \text{ aqueous salt concentration} = \frac{\text{NaCl content}}{\text{NaCl content} + \text{moisture content}} \times 100$$

NaCl content = grams of NaCl (salt) in 100 g of product.
Moisture content = grams of water in 100 g of product.

Use of different solutes often results in different growth-limiting water activity effects on organisms; when glycerol is used to reduce water activity, the growth-limiting water activity for *C. botulinum* type E is 0.94, but using salt it is 0.97 (Table 3.6). This could indicate that the growth-inhibitory effect of salt is not entirely due to lowering of water activity. In foods, the most commonly used agent affecting water activity is salt but some other solutes may also be present and make some additional contribution to the reduction of water activity, e.g. sugars. Other agents may increase water activity limits, e.g. preservatives (Hauschild, 1989). Because of the complexity of the physico-chemistry of many foods, in those foods for which water activity is an important *C. botulinum* growth-controlling factor, routine monitoring by measurement of water activity is often recommended. In foods where water activity measurement is either unreliable, e.g. fatty foods, or where it varies only slightly, e.g. processed cheese, the measurement of aqueous salt concentration can be more useful and is often preferred.

The combined effect of non-optimal conditions of temperature, water activity and pH on the growth and survival of microorganisms usually has a greater effect than an individual factor used at the same level (International Commission on Microbiological Specifications for Foods, 1996). Preservatives and packaging atmosphere can also be used in combination with these factors to enhance microbial suppression. The International Commission on Microbiological Specifications for Foods (1980 and 1996) gives extensive information on each of the above factors including the effect of growth-inhibitory interactions and the production of botulinum toxin. Table 3.9 shows some selected data indicating the time to *C. botulinum* toxin production found in a variety of food based studies in which pH, salt concentration and storage temperature are the key controlling factors. Clearly, the lower the pH, the higher the salt concentration and the lower the storage temperature, the longer the time taken to produce toxin. In real foods, all the potential controlling factors must be balanced for organoleptic, specific food character and commercial considerations as well as for food safety reasons.

An important group of food products for which combinations of inhibitory factors are essential is that of hams and bacon and similarly cured meats in which salt concentration, concentration of sodium and/or potassium salts of nitrite and nitrate, polyphosphates, sugars and sometimes other ingredients combine to create a specific food character with commercially desirable organoleptic properties.

The extent of inhibition of the growth of *C. botulinum* by sodium nitrite is

Table 3.9 Time taken to toxin production by non-proteolytic *C. botulinum* in foods with different pH levels and salt concentrations when stored at different temperatures, adapted from Graham *et al.* (1996a)

Food type	Botulinum type	pH	NaCl (%)	Storage temperature (°C)	Time to toxin production (days)
Crabmeat homogenate (sterilised)	E	5.4	1.25	10	9
Turkey (cooked and vacuum packed)	B and E	6.3 6.3	1.47 2.25	16 8	2 9
Beef (homogenised and sterilised)	B, E and F	6.2 6.2	1.0 1.0	12 6	2 7
Ham (cured)	B and E	5.8	0.4	8	11
Beef (sous-vide)	B and E	5.8	0.96	8	8
Chicken (sous-vide)	B and E	5.9	0.52	8	16

affected by salt concentration, pH and temperature although there is some debate over whether in-going or residual nitrite levels are important (Lücke and Roberts, 1993). Table 3.10 indicates the effect of interactions of salt, sodium nitrite and temperature on toxin production by different types of *C. botulinum*. Nitrite is depleted over time and depletes more rapidly with decreasing pH and/or increasing temperature (Kim and Foegeding, 1993). The presence of low levels of isoascorbate also enhances the anti-botulinal effect of nitrite probably by sequestering any iron present in meat, the latter reducing the efficacy of nitrite inhibition (Lücke and Roberts, 1993) (see section on 'Refrigerated, cooked, cured and uncured meat products' in Chapter 4).

Health concerns related to the use of nitrite, particularly in meat curing processes, have arisen because of the formation of potentially carcino-genic and mutagenic nitrosamines from the reaction between nitrous acid (from hydration of nitric oxide produced from the reduction of sodium nitrite) and secondary amines present in the food. Some bacteria including clostridia may also produce nitrosamines from nitrite (Kim and Foegeding, 1993). This concern led to the investigation of alternative approaches to the use of nitrite in meat curing processes but, to date, no substitute has

Table 3.10 Time taken to toxin production by *C. botulinum* in foods with different salt and sodium nitrite concentrations when stored at different temperatures, adapted from International Commission on Microbiological Specifications for Foods (1996)

Food type	Botulinum type	NaCl (% aqueous salt)	Sodium nitrite (ppm : mg/kg)	Storage temperature (°C)	Time to toxin production (days)
Cooked ham, sliced and vacuum-packed	A and proteolytic B	3.1	88	15	>30
Cooked ham, sliced and vacuum-packed	A and proteolytic B	3.1	88	25	2-5
Cooked ham, sliced and vacuum-packed	A and proteolytic B	4.1	0	20	15
			50		25
			100		>25
			200		30
			0	25	3
			50		4
			100		5
			200		8
			0	30	2
			50		3
			100		3
			200		8
Pork tongue, jellied and vacuum-packed	A and proteolytic B	4.3	75	20	30
			150		20
			75	25	10
			150		10
			75	30	6
			150		5

been found that gives the same or similar organoleptic characteristics and contribution to microbial inhibition as that produced by the use of nitrite.

The research carried out has nonetheless contributed an understanding of the *C. botulinum* growth-inhibiting activity of some other compounds often applied in combination with low levels of sodium nitrite, e.g. sorbic acid and salts of sorbic acid, polyphosphates, particularly sodium acid pyrophosphate and diphosphates (Kim and Foegeding, 1993; International Commission on Microbiological Specifications for Foods, 1996) and sodium lactate (Houtsma *et al.*, 1994; International Commission on Microbiological Specifications for Foods, 1996). Also, a considerable amount of work has been carried out on the use of nisin as an anti-clostridial agent particularly in dairy products, e.g. processed cheese; this is briefly reviewed by Kim and Foegeding (1993).

National and international legislation places limits on the levels of many preservative compounds in specific foods and local legislation must be consulted for further information, e.g. in UK legislation, *The Miscellaneous Food Additives Regulations, 1995, Statutory Instrument No. 3187*, HMSO, London. In addition to these, the potential inhibitory effect of some herbs and spices and their oils on the spore germination and cell growth of *C. botulinum* has been studied with varied results (Kim and Foegeding, 1993). The inhibitory effect of these materials is not considered sufficiently reliable for use as a sole means for controlling the growth of *C. botulinum*.

Although the effect of some other organisms on growth and toxin production by *C. botulinum* has already been referred to above, other bacteriocin-producing organisms such as *Lactobacillus* spp. and *Bacillus* spp. can have a direct inhibitory effect on the growth of *C. botulinum* or may indirectly prevent potential cases of botulism by causing overt food spoilage such that a consumer would reject the food for consumption (Hauschild, 1989, Kim and Foegeding, 1993).

The effect of ionizing radiation on the spores of *C. botulinum* has been studied. The spores of proteolytic strains have been found to have *D* values in the range 2.0–4.5 kGy at temperatures between –50°C and –10°C and type E strains 1.0–2.0 kGy. Sensitivity to ionizing radiation is reported to be affected by the presence of oxygen, preservatives and temperature (Hauschild, 1989, Kim and Foegeding, 1993). *C. botulinum* toxin is not inactivated by ionizing radiation levels likely to be applied to foods (International Commission on Microbiological Specifications for Foods, 1996).

A considerable amount of work has been done in recent years to examine the effect of combinations of heat treatments, pH, salt concentration and other factors on the growth from spores of different types of *C. botulinum*. Most of this work, however, has been carried out using artificial growth media (Graham *et al.*, 1996a and b, Graham *et al.*, 1997, Stringer and Peck, 1997) or simulated food conditions (Dodds, 1989, Brown *et al.*, 1991, Ter Steeg *et al.*, 1995, Carlin and Peck, 1996) and the results cannot relate easily or reliably to real foods, the complexity and diversity of which are great, and in which apparent safety factors may be nullified by some other aspect of product formulation, packaging, etc. Despite this, the results from such work are used to create mathematical models for the growth of and toxin production by *C. botulinum* in foods with different combinations of selected inhibitory factors and the models are used by some to predict growth/no growth in products of these varied formulations. The use of such models and the interpretation of results need careful consideration and it is not advisable for such models to be used by non-microbiologists or microbiologists who have no direct food industry knowledge and experience.

Because foods are so complex in respect of their chemical, biochemical and physical nature as well as the external factors of packaging type, packaging atmosphere and temperature, there is often more than just a single factor involved in the control of *C. botulinum* growth and toxin production in a food. Small changes to product recipes, e.g. reduction in salt level, use of alternative sweeteners (aspartame instead of sucrose), packaging atmosphere or storage temperature could have a significant and adverse effect on the safety of a food with respect to *C. botulinum*. Physico-chemical factors combine in food to form micro-environments and environmental gradients (temperature, pH, salt, etc.) may form and, depending on the conditions of the immediate environment, a spore of *C. botulinum*, if present, may be inhibited or may germinate and grow to produce toxin. It is for these reasons that challenge studies using real foods provide the most useful information about the potential safety and stability of a specific food. Unfortunately, such studies are often only carried out with a specific food after an outbreak has occurred implicating that food type (Table 3.11). Most of these studies are more specifically detailed in the relevant product categories in Chapter 4 of this book.

Botulism alerts are still in evidence and involve a variety of foods, e.g. black olives, pickled vegetables in oil and smoked fish in one food safety alert (Anon, 1999). It is essential that food microbiologists and technologists understand the physico-chemical properties of the particular foods and processes for which they have responsibility and the need to imple-

Table 3.11 Challenge studies carried out on the growth of *C. botulinum* in real food formulations

C. botulinum type and inoculum level	Food	Factors/conditions	Reference
Proteolytic A < 10–4000/g	Sautéed onions	Restaurant recipe, process and containers for sautéed onions incubated at 35°C	Solomon and Kautter (1986)
Naturally occurring spores in fresh fish	27 types of fresh fish	Vacuum packaged and stored at 12°C	Lilly and Kautter (1990)
Proteolytic A & B	Fresh pasta; flat noodle and filled tortellini	Water activity 0.92–0.99 Modified atmosphere packaged Stored at 4°C and 30°C	Glass and Doyle (1991)
Non-proteolytic E 10^3–10^4/g	Rainbow trout	Vacuum packaged in oxygen barrier film, stored at 4°C and 10°C. Vacuum packaged in oxygen permeable film, stored at 4°C	Garren *et al.* (1995)
Proteolytic A & B Non-proteolytic B 100 spores/g	Romaine (cos) lettuce Shredded cabbage	Two packaging types used (vented and non-vented) Incubated at 4.4°C, 12.7°C or 21°C	Petran *et al.* (1995)
Proteolytic A & B 1×10^3 spores/g	Sous-vide spaghetti and meat sauce	pH 4.5–6.0 a_w 0.992–0.972 Vacuum packaged Heat process 75°C for 36 minutes Storage at 15°C	Simpson *et al.* (1995)

Table 3.11 Continued

C. botulinum type and inoculum level	Food	Factors/conditions	Reference
Proteolytic A & B Non-proteolytic B & E 100 spores/g	Fresh prepared vegetables: coleslaw mix (chopped cabbage and carrots), broccoli florets, sliced carrots, green beans, shredded lettuce	Vacuum or air packaged in different packaging types (O_2/ CO_2 transmission rates) and incubated at 4°C, 12°C or 21°C	Larson et al. (1997)
Non-proteolytic E	Fresh salmon fillet steaks	Vacuum or modified atmosphere packaged (75% CO_2 : 25% N_2) or 100% air Storage at 4°C, 8°C or 16°C	Reddy et al. (1997)
Non-proteolytic B, E & F Approx. 10^2–10^3 or 4×10^3–2×10^4/cfu/kg (high and low inocula)	Cold-smoked rainbow trout	Sodium nitrite and potassium nitrate in vacuum packed fish stored at 4°C and 8°C	Hyytiä et al. (1997)

ment measures to prevent the hazard of *C. botulinum* from becoming an outbreak of botulism. Properly structured hazard analysis based on such an understanding and supported, as appropriate, by predictive mathematical modelling information and, where indicated by the hazard analysis, by relevant product challenge test data will help maintain the food industry's safety record in respect of this potentially lethal pathogen.

4

INDUSTRY FOCUS: CONTROL OF
CLOSTRIDIUM BOTULINUM

INTRODUCTION

The significant number of outbreaks of botulism throughout the world clearly demonstrates the necessity for the food industry to employ measures to control this most serious of hazards. Although different product types can present a risk of causing outbreaks of botulism, in many outbreaks it is a failure in the control systems that can be identified as a major contributory factor in the realisation of the risk. In such cases, the application of a hazard analysis approach and consequent implementation of controls at the critical points could have prevented the outbreaks, provided the control systems were operated consistently correctly. It is strongly recommended that all persons involved in the production, processing and sale of food adopt a hazard analysis approach considering all relevant pathogens, including *C. botulinum*. Indeed, the requirement to operate such an approach in food businesses is embodied in European and national laws.

To help focus attention on the products representing the greatest concern in relation to *C. botulinum* and, consequently, the areas requiring greatest management control, a series of questions can be applied to each food process/product (Table 4.1). Processes and products can be reviewed against the key questions to identify the level of concern that the organism may represent. As a guide to answering these questions, some familiar products in different commodity groups are given as examples in Table 4.2.

After answering each of the questions in Table 4.1, the product can be assessed against the profiles given in Table 4.3 to determine the level of concern that may be associated with the product. Having done this, the key process areas requiring greatest attention for control of the hazard can be determined (Table 4.4).

Table 4.1   How much of a concern does your product represent?

Question	Yes	No
Is *C. botulinum* expected to be present in the raw material?	✓	
Will the spores of *C. botulinum* be destroyed or reduced to an acceptable level by any of the processing stages e.g. 90°C for 10 minutes for non-proteolytic strains and 121°C for 3 minutes for proteolytic strains?		
Is the product at risk from post-process contamination?		
Could the process allow the growth of *C. botulinum* (not necessarily to toxin forming levels)?		
Will the final product formulation allow the growth of *C. botulinum*?		
Does the product have an extended shelf life, i.e. > 10 days?		
Are storage restrictions, i.e. chilling, and shelf life necessary or employed to prevent growth of *C. botulinum*?		

Every process and product will differ from those presented in the tables; therefore, the tables should be used for guidance purposes only. Complete understanding of the hazard and necessary controls can only be gained by applying a full hazard analysis to each production process. In addition, it is important to note that even processes and products which are rated as being of very low concern in relation to *C. botulinum* may still be capable of causing outbreaks if the controls inherent in the normal manufacture of these products are not applied consistently correctly. In fact, significant hazards to food safety are presented by complacent management teams who believe that their product is safe because of historical precedence or from a food production team lacking the necessary skills and training in safe food manufacture. Food products are most often made safe or unsafe to eat by human intervention.

The highest concern products with regard to *C. botulinum* are those where the organism may be present, even in low numbers, in the raw material, where the process does not eliminate it, due to design or fault, and where it may grow during the process or in the finished product which is consumed without any further processing. Products such as non-preserved vegetables in oil and minimally processed or lightly 'fermented' meat and fish would fall within this category and, indeed, it is not surprising that such products have been implicated in outbreaks of botulism.

Table 4.2 Examples of key process stages where *C. botulinum* may represent a hazard in different foods

Product	Product examples	Raw material contamination	Destruction process (NP/P)	Post-process contamination	Process allows growth*	Product formulation allows growth (NP/P)	Extended shelf life	Storage restrictions required to prevent growth
Dairy products								
Soft ripened cheese	Brie, Camembert	Yes	No/No	Yes	No	Yes†/Yes†	Yes	Yes
Hard cheese	Cheddar, Parmesan	Yes	No/No	Yes	No	No/No	Yes	No
Acid, fermented milk products (pH <4.7)	Yoghurt, clotted cream	Yes	No/No	Yes	No	No/No	Yes	No
Low acid, fermented/acidified milk products (pH >5)	Mascarpone, cream cheese	Yes	No/No	Yes	No	Yes/Yes	Yes	Yes
Processed cheese – chilled storage	Cheese spread	Yes	Yes†/No	Yes†	No	No/Yes	Yes	Yes
– ambient storage	Processed cheese	Yes	Yes†/No	Yes†	No	No/No†	Yes	No
Heat processed, extended life, chilled dessert	Mousse	Yes	Yes†/No	Yes†	No	Yes†/Yes	Yes	Yes
Meat products								
Raw meat and poultry – vacuum/MAP	Beef, pork, lamb, chicken	Yes	No/No	Yes	No	Yes/Yes	No†	Yes
Fermented meat	Salami	Yes	No/No	Yes	No†	No/No†	Yes	No†
Dry cured meat	Parma ham	Yes	No/No	Yes	Yes†	No/No†	Yes	No†

See p. 91 for key to table.

Table 4.2 Continued

Product	Product examples	Raw material contamination	Destruction process (NP/P)	Post-process contamination	Process allows growth*	Product formulation allows growth (NP/P)	Extended shelf life	Storage restrictions required to prevent growth
Cooked ($<90°C$) cured meat - vacuum/MAP	Ham, pâté, frankfurter	Yes	No/No	Yes	No	Yes†/Yes†	Yes	Yes
Cooked ($<90°C$) uncured meat - vacuum/MAP	Roast pork, chicken	Yes	No/No	Yes	No	Yes†/Yes	Yes	Yes
Canned, cured meats ($F_o > 0.5 - <3$)	Luncheon meat, canned ham	Yes	Yes/Yes†	No	No	No/No†	Yes	No†
Fish and shellfish								
Raw fish and shellfish – vacuum/MAP (to be cooked)	Cod, mussels	Yes	No/No	Yes	No	Yes/Yes	No†	Yes
Raw fish and shellfish, consumed raw	Oysters, sushi	Yes	No/No	Yes	No	Yes/Yes	No	Yes
Cold smoked fish – vacuum/MAP (consumed raw)	Smoked salmon, smoked trout	Yes	No/No	Yes	Yes†	No†/Yes	Yes	Yes
Cooked ($<90°C$) fish and shellfish - vacuum/MAP	Prawns, crab	Yes	No/No	Yes	No	Yes†/Yes	No†	Yes
Canned fish ($F_o > 3$)	Canned tuna, salmon, fish paste	Yes	Yes/Yes	No	No	Yes/Yes	Yes	No

See p. 91 for key to table.

Table 4.2 Continued

Product	Product examples	Raw material contamination	Destruction process (NP/P)	Post-process contamination	Process allows growth*	Product formulation allows growth (NP/P)	Extended shelf life	Storage restrictions required to prevent growth
Salads, vegetables and fruit								
Raw salads and vegetables (loose)	Potatoes, peas, lettuce, celery	Yes	No/No	Yes	No	No†/No†	No†	No†
Prepared, MAP salads and vegetables	Salad mix, broccoli	Yes	No/No	Yes	No	Yes/Yes	No†	Yes
Prepared vegetables and salads (non-MAP)	Salad cress, sliced and shredded lettuce	Yes	No/No	Yes	No	Yes/Yes	No	Yes
Vegetables in oil – ambient stored	Garlic in oil	Yes	Yes†/No	Yes	No	Yes†/Yes†	Yes	No
Canned vegetables ($F_o > 3$)	Canned beans, peas	Yes	Yes/Yes	No	No	Yes/Yes	Yes	No
Cooked vegetables	Baked potato	Yes	Yes/No	Yes	No	Yes/Yes	No	Yes
Dried fruit (high acid)	Figs, tomatoes	Yes	No/No	Yes	No†	No/No	Yes	No
Canned, high acid fruit	Jam, figs in syrup	Yes	Yes†/No	No	No	No/No†	Yes	No

*: Under conditions of good manufacturing practice.

†: Some processes, formulations or conditions will differ which may require the contrary response.

NP: Non-proteolytic.

P: Proteolytic.

MAP: Modified Atmosphere Packed.

Note: Individual products and processes can vary significantly and due account must be taken of differences in formulation and process parameters when viewing this table.

Raw material contamination: Is *C. botulinum* expected to be present in the raw material?

Destruction process: Will the spores of *C. botulinum* be destroyed or reduced to an acceptable level by any of the processing stages, e.g. 90°C for 10 minutes for non-proteolytic strains and 121°C for 3 minutes for proteolytic strains?

Post-process contamination: Is the product at risk from post-process contamination?

Process allows growth: Could the process allow the growth of *C. botulinum* (not necessarily to toxin forming levels)?

Product formulation allows growth: Can the final product formulation and conditions allow the growth of *C. botulinum*?

Extended shelf life: Does the product have an extended shelf life, i.e. > 10 days?

Storage restrictions required to prevent growth: Are storage restrictions, i.e. chilling and shelf life, necessary or employed in the distribution or retail systems, and/or by the consumer to prevent growth of *C. botulinum*?

Information given is for guidance only and may not be appropriate for individual circumstances. It is recommended that proper hazard analysis is carried out for every process and product to identify where controls must be implemented to minimise the hazard from *C. botulinum*.

Table 4.3 Categories of concern

Level of concern	Product examples	Raw material contamination	Destruction process (NP/P)	Post-process contamination	Process allows growth*	Product formulation allows growth (NP/P)	Extended shelf life	Storage restrictions required to prevent growth
Category 1: Highest	Vegetables in oil – ambient stored	Yes	Yes†/No	Yes	No	Yes†/Yes†	Yes	No
	Cooked (<90°C) uncured meat – vacuum/MAP	Yes	No/No	Yes	No	Yes†/Yes	Yes	Yes
	Cold smoked fish – (vacuum/MAP) consumed raw	Yes	No/No	Yes	Yes†	No†/Yes	Yes	Yes
	Low acid, fermented/ acidified milk products (pH > 5)	Yes	No/No	Yes	No	Yes/Yes	Yes	Yes
Category 2: High	Soft ripened cheese	Yes	No/No	Yes	No	Yes†/Yes†	Yes	Yes
	Heat processed, extended life, chilled dairy dessert	Yes	Yes†/No	Yes†	No	Yes†/Yes	Yes	Yes
	Cooked (<90°C) cured meat – vacuum/MAP	Yes	No/No	Yes	No	Yes†/Yes†	Yes	Yes
	Cooked (<90°C) fish and shellfish – vacuum/MAP	Yes	No/No	Yes	No	Yes†/Yes	No†	Yes

See p. 94 for key to table.

Table 4.3 Continued

Level of concern	Product examples	Raw material contamination	Destruction process (NP/P)	Post-process contamination	Process allows growth*	Product formulation allows growth (NP/P)	Extended shelf life	Storage restrictions required to prevent growth
Category 3: Medium	Raw meat and poultry – vacuum/MAP	Yes	No/No	Yes	No	Yes/Yes	No†	Yes
	Raw fish and shellfish – vacuum/MAP (to be cooked)	Yes	No/No	Yes	No	Yes/Yes	No†	Yes
	Fermented meat	Yes	No/No	Yes	No†	No/No†	Yes	No†
	Dry cured meat	Yes	No/No	Yes	Yes†	No/No†	Yes	No†
	Processed cheese – chilled storage	Yes	Yes†/No	Yes†	No	No/Yes	Yes	Yes
	Prepared, MAP salads and vegetables	Yes	No/No	Yes	No	Yes/Yes	No†	Yes
	Cooked vegetables	Yes	Yes/No	Yes	No	Yes/Yes	No	Yes
Category 4: Low	Hard cheese	Yes	No/No	Yes	No	No/No	Yes	No
	Raw salads and vegetables (loose)	Yes	No/No	Yes	No	No†/No†	No†	No†
	Acid, fermented milk products (pH <4.7)	Yes	No/No	Yes	No	No/No	Yes	No

See p. 94 for key to table.

Table 4.3 Continued

Level of concern	Product examples	Raw material contamination	Destruction process (NP/P)	Post-process contamination	Process allows growth*	Product formulation allows growth (NP/P)	Extended shelf life	Storage restrictions required to prevent growth
Category 4: Low	Canned, cured meat ($F_o > 0.5 - < 3$)	Yes	Yes/Yes†	No	No	No/No†	Yes	No†
	Canned meat, fish and vegetables ($F_o > 3$)	Yes	Yes/Yes	No	No	Yes/Yes	Yes	No
Category 5: Lowest	Dried fruit (high acid)	Yes	No/No	Yes	No†	No/No	Yes	No
	Canned, high acid fruit	Yes	Yes†/No	No	No	No/No†	Yes	No

*: Under conditions of good manufacturing practice.

†: Some processes, formulations or conditions will differ which may require the contrary response.

NP: Non-proteolytic.

P: Proteolytic.

MAP: Modified Atmosphere Packed.

Highest/high concern: A product where no destruction process is applied and where *C. botulinum* could be present due to raw material contamination or as a post-process contaminant **and** where the process allows growth or, the final product allows growth over an extended shelf life, especially where reliance is placed on distribution and consumer chilled storage to maintain safety.

Medium concern: Where *C. botulinum* may be present in the raw material or as a post-process contaminant **and** where the process or final product may allow growth although this may be restricted by an inhibitory formulation or short product shelf life.

Lowest/low concern: Where *C. botulinum* may be present in the raw material but either the process applied destroys the organism and it cannot recontaminate the product or, if it may be present in the final product, the product formulation reliably prevents growth. Information given is for guidance only and may not be appropriate for individual circumstances. It is recommended that proper hazard analysis is carried out for every process and product to identify where controls must be implemented to minimise the hazard from *C. botulinum*.

Table 4.4 Stages where control of *C. botulinum* is critical (based on the categories of concern)

Level of concern	Product examples	Raw material control	Destruction process	Post-process contamination	Process conditions	Product formulation	Distribution/retail/ consumer issues
Category 1: Highest	Vegetables in oil – ambient stored	Yes		Yes		Yes	Yes
	Cooked (<90°C) uncured meat – vacuum/MAP	Yes		Yes	Yes	Yes	Yes
	Cold smoked fish – (vacuum/MAP) consumed raw	Yes		Yes	Yes	Yes	Yes
	Low acid, fermented/ acidified milk products (pH > 5)	Yes			Yes		Yes
Category 2: High	Soft ripened cheese	Yes		Yes	Yes		Yes
	Heat processed, extended life, chilled dairy dessert	Yes	Yes	Yes		Yes	Yes
	Cooked (<90°C) cured meat – vacuum/ MAP	Yes		Yes	Yes	Yes	Yes
	Cooked (<90°C) fish and shellfish – vacuum/MAP	Yes		Yes	Yes		Yes

See p. 97 for key to table.

Table 4.4 Continued

Level of concern	Product examples	Raw material control	Destruction process	Post-process contamination	Process conditions	Product formulation	Distribution/retail/consumer issues
Category 3: Medium	Raw meat and poultry – vacuum/MAP	Yes					Yes
	Raw fish and shellfish – vacuum/MAP (to be cooked)	Yes					Yes
	Fermented meat	Yes			Yes	Yes	
	Dry cured meat	Yes			Yes	Yes	
	Processed cheese – chilled	Yes	Yes			Yes	Yes
	Prepared, MAP salads and vegetables	Yes		Yes			Yes
	Cooked vegetables	Yes		Yes			Yes
Category 4: Low	Hard cheese	Yes			Yes	Yes	Yes
	Raw salads and vegetables (loose)	Yes					
	Acid, fermented milk products (pH <4.7)	Yes			Yes	Yes	
	Canned, cured meat ($F_o > 0.5 - <3$)	Yes	Yes	Yes	Yes	Yes	
	Canned meat, fish and vegetables ($F_o > 3$)	Yes	Yes	Yes			

Table 4.4 Continued

Level of concern	Product examples	Raw material control	Destruction process	Post-process contamination	Process conditions	Product formulation	Distribution/retail/consumer issues
Category 5: Lowest	Dried fruit (high acid)	Yes		Yes	Yes	Yes	
	Canned, high acid fruit	Yes	Yes	Yes	Yes	Yes	

MAP: Modified Atmosphere Packed.

Highest/high concern: A product where no destruction process is applied and where *C. botulinum* could be present due to raw material contamination or as a post-process contaminant **and** where the process or the final product allows growth over an extended shelf life, especially where reliance is placed on distribution and consumer chilled storage to maintain safety.

Medium concern: Where *C. botulinum* may be present in the raw material or as a post-process contaminant **and** where the process or final product may allow growth although this may be restricted by an inhibitory formulation or short product shelf life.

Lowest/low concern: Where *C. botulinum* may be present in the raw material but either the process applied destroys the organism and it cannot recontaminate the product or, if it may be present in the final product, the product formulation reliably prevents growth.
Information given is for guidance only and may not be appropriate for individual circumstances. It is recommended that proper hazard analysis is carried out for every process and product to identify where controls must be implemented to minimise the hazard from *C. botulinum*.

REFRIGERATED, COOKED, CURED AND UNCURED MEAT PRODUCTS

Cooked meat products comprise a large group of diverse products which constitute a significant part of the diet of many individuals in the developed world. Products include cooked ham, chicken and beef, together with cooked comminuted meat such as pâté. They may be cooked and sold as whole joints although most tend to receive some further processing, e.g. slicing by the manufacturer or retailer. These products are also significantly represented in composite foods such as prepared meals. Cooked, chilled meats have an enviable safety record with respect to *C. botulinum,* with few recorded outbreaks of botulism, which is surprising given the likely presence of *C. botulinum* and the mild processing that these products receive. The products are sold as ready-to-eat and are therefore frequently consumed cold, with no further processing by the consumer.

Description of process

The process of manufacture of these products is fairly straightforward (Figure 4.1) and involves cooking bulk packed meat as whole joints or as part of a comminuted mix formed into blocks or sausage shapes of varying sizes. The meat is subjected to an extended cook which, in the UK, is designed to ensure that all parts of the product achieve a minimum heat process of 70°C for 2 minutes or an equivalent heat process. The heat process can vary significantly for different products and is usually guided by the desire to achieve an organoleptically acceptable finished product with minimum reduction in weight due to water loss.

Cooking usually takes place using ovens built into a dividing wall; a double door system in the oven allows the raw meat to be placed in from one side and, after heat processing, the cooked meat is removed from the other side, thereby preventing the hazard of contamination of cooked meat from raw meat and raw meat handling equipment and personnel. Cooked meat products are then usually blast-chilled to reduce the temperature rapidly to 0–5°C in an attempt to prevent the germination and growth of any surviving spore-forming bacteria.

In many cases, cooked bulk meats in hermetically sealed containers may be supplied to a distributor or retailer without further processing and are therefore not subject to post-process contamination prior to receipt and handling by the retailer. Some meats are cooked in cans at 'pasteurisation' temperatures/times which then require subsequent chilled storage to maintain safety and stability. These products are subject to the same potential hazards with regard to *C. botulinum* as meats cooked in other forms of packaging. However, it is common practice for many meats to receive some form of secondary processing by the manufacturer, which in its simplest form may involve removing the outer packing and re-cooking to achieve a roasted appearance. Alternatively, they may be sliced into smaller blocks for display on delicatessen counters or into small packs for sale as pre-pack units.

The shelf lives of cooked meats vary significantly depending on the cooked meat type and formulation and the degree of post-process handling and contamination that they are exposed to. In general, uncured cooked meats, such as chicken, pork and beef, may be allocated shelf lives varying from less than 10 days to 20 days or more. Cured meats including many pâtés and hams and canned, cured, chilled meats may be given shelf lives ranging from 15 days to several months. Shelf lives of those cured products that are cooked in their final container are usually much longer as the potential hazard associated with the presence and growth of *Listeria monocytogenes* is eliminated and the hazard of non-proteolytic *C. botulinum* is controlled by the formulation. Cooked, sliced meats, on the other hand, usually have a restricted life both due to the need to ensure that any contaminants like *L. monocytogenes* are not afforded extended shelf lives in which to grow and also because post-process contaminants including lactic acid bacterial types will inevitably cause spoilage.

Raw material issues and control

Raw materials used to produce cooked meat products can vary considerably but the major component is the meat itself, which may be from a number of animal species including chicken, pork, beef and lamb, used singly or in combination. Increasingly, the developed world is also being introduced to exotic varieties of meat including ostrich, alligator and kangaroo, although these have formed part of the diets of the indigenous populations of the countries where they originate for many centuries.

Raw meats are often treated with salt, spices, herbs (often as extracts) and emulsifying agents (phosphates and polyphosphates) added by injection of brine (water and sodium chloride) and/or they may be added by tumbling meat with dry salt/spice mixtures or immersing meat in brine tanks for several hours or days under refrigerated conditions.

Products which are made from comminuted mixes such as pâté, frankfurters and many sliced meats are manufactured by bowl chopping the meat raw materials, then adding and mixing salt and other ingredients prior to stuffing the mix into casings. The raw meat for many of these

Process Stage	Consideration
Animal husbandry ↓	Health Cleanliness
Animal slaughter and processing ↓	Hygiene Temperature
Meat transport, delivery and storage ↓	Hygiene Temperature
Comminuted and reformed bulk meats Bowl chopping and addition of other ingredients (spices, herbs, salt, etc.) or	Hygiene Temperature Distribution of preservative factors
Whole joints of meat Brine injection (where applicable, e.g. hams and cured meats) ↓	Hygiene Temperature Distribution of preservative factors
Cooking ↓	Temperature Time High/low risk segregation (post cooking)
Blast chilling ↓	Hygiene Temperature Time
Removal from container (where applicable) ↓	Hygiene
Storage ↓	Hygiene Temperature Time
Roasting/chilling, where applicable ↓	Temperature Time
Super chilling (0–2°C) ↓	Hygiene
Slicing, where applicable ↓	Hygiene
Garnishing, where applicable ↓	Raw material control Hygiene

Figure 4.1 Process flow diagram and technical considerations for a typical cooked meat product.

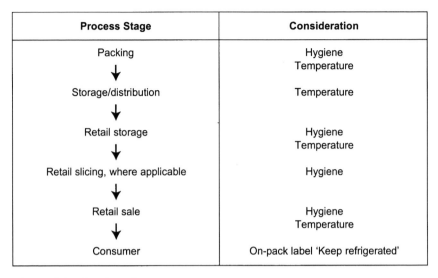

Process Stage	Consideration
Packing	Hygiene Temperature
Storage/distribution	Temperature
Retail storage	Hygiene Temperature
Retail slicing, where applicable	Hygiene
Retail sale	Hygiene Temperature
Consumer	On-pack label 'Keep refrigerated'

Figure 4.1 Continued

comminuted products may consist of less prime material including liver and other tissues together with the addition of fat; these can add significantly to the initial microbial load of the primary mix as these materials may carry higher levels of microbial contaminants than fresh whole muscle tissue.

Many cooked meat products, particularly comminuted meats, have the preservative agent sodium nitrite added, which may also be supplemented with sodium nitrate. Sodium nitrite contributes flavour to the product but is also a critical factor in preventing the growth of contaminating clostridia. Cooked meat products made from pork often have added nitrite, i.e. hams, whereas poultry and beef and other meat species are rarely preserved by the use of nitrite.

Because of the wide variety of raw materials used to manufacture these products, it is clear that the microflora will be significantly varied in number and type of bacteria present. *C. botulinum* will be present on occasion in these materials and the likelihood of this occurring will depend on the specific raw materials used and their origin. Different types of *C. botulinum* are known to be more dominant in different parts of the world and in different habitats and this has been extensively reviewed by Hauschild (1989). In general, if the raw material is from an aquatic habitat the *C. botulinum* type is more likely to be type E whereas those from terrestrial sources will be more commonly contaminated with types A and B (Table 1.7). Most surveys for *C. botulinum* have been conducted on soil and aquatic sediment samples rather than food items, with results showing incidences of *C. botulinum* ranging from 0% to over 90% and levels from < 1 to over 2000 spores/kg of sample tested in both environments. Therefore, it should be assumed by all processors of food raw materials that, on occasion, the material will harbour spores or vegetative cells of one, if not a number of *C. botulinum* types. The systems of control in the manufacture of the food must therefore take this hazard into account.

Surveys of raw and processed meats have shown a variable incidence of *C. botulinum* (Table 4.5) with percentage incidence ranging from 0% to 15% in raw pork, c. 1% in cured meats and just over 1% in liver sausage (Hauschild, 1989). Levels of contamination were generally very low (< 10 spores/kg).

In addition to the raw meats, it is most likely that any herbs and spices present in cooked comminuted meat products such as pâté will also contribute to the spore loading. The growing and processing conditions that these raw materials undergo in the countries of origin may expose them to a significant amount of soil and soil dust which is known to be heavily laden with microbial contaminants, spore-formers often being the predominant types. The levels of *Bacillus* spp. and *Clostridium* spp.

Table 4.5   Incidence of *C. botulinum* in meat products, adapted from Hauschild (1989)

Source	Food	Incidence (percentage)	Number per kg (MPN)	Types found
North America	Raw meat	1/2358 (0.04)	0.1	C
	Cured meats	1/90 (1.11) 6/372 (1.61) } 2/132 (1.52)	0.2–0.6	A, B
	Bacon	1/208 (0.48)	0.1	NR
	Liver sausage	5/276 (1.81)	0.2	A
United Kingdom	Raw pork	0/280 (0) 3/140 (2.14) 20/138 (14.49) } 7/126 (5.56)	<0.1–5	NR
	Bacon	11/263 (4.18) 6/110 (5.45) } 19/26 (73.08)	1–7	A, B

NR = not recorded

found in untreated dried herbs and spices may be in excess of $10^2/g$ and can frequently exceed $10^4/g$. From time to time, a proportion of these will undoubtedly be *C. botulinum*. Dodds (1993b) reported a survey of produce and fresh herbs where *C. botulinum* was found in dill and parsley. Many processors use spice extracts to improve consistency of flavour and quality and such material is not usually of concern in relation to the presence of microbial contaminants.

It is likely to be impossible to achieve raw material sources consistently free from contamination by such ubiquitous organisms as spore-forming bacteria and, while it may be prudent to source, store and process raw materials in a way that will preclude the germination and growth of these contaminants, it is inevitable that they, and *C. botulinum*, will be present on occasion.

Therefore, the most important consideration in relation to these and other raw materials such as the meat components is the use of appropriate storage conditions. Perishable products should be stored under conditions that will prevent the growth of any contaminants, using either freezing or chill conditions and dry raw materials such as herbs and spices should be maintained in cool, dry, pest-free storage areas. For raw materials such as dried herbs and spices it may be useful to monitor the spore loading of incoming batches, with high levels being reported to the raw material supplier for action to improve quality to ensure that consistently poor materials are not received.

Process issues and control

It is important in the control of any microorganism to limit the spread of the organism and the levels at which it may be present. Therefore, while it may appear only of cosmetic benefit in relation to the control of *C. botulinum* in cooked meat, it is still important to ensure that processing equipment and practices associated with the handling and preparation of raw meat and meat mixes are operated to minimise cross-contamination to successive batches. Limiting the occurrence and levels of any hazard has the effect of reducing the overall risk associated with the potential for that hazard to be exposed to the general population, even though in the case of *C. botulinum* the organism must also germinate, grow and produce toxin before it can cause illness. Operation of effective cleaning practices of common food contact machinery such as bowl chopping machines, mixers and fillers together with maintenance of environmental hygiene can assist in reducing the chances of passing any contaminants from one batch of raw materials to successive batches.

In products such as some cooked, cured meats which rely on the formulation of the finished product to prevent the growth of *C. botulinum*, it is essential that agents acting to prevent or restrict growth are distributed evenly during the raw material processing stage. This would naturally apply to the injection of brine and the addition of curing salts including sodium nitrite. Ingoing levels must be carefully monitored and equilibration of the preservative agents throughout the meat product is equally important. This is more difficult to achieve by product immersion than by injection systems or comminution and mixing, and this should be taken into account when processing products. Monitoring the level of anti-botulinum factors such as aqueous salt and nitrite is usually an important process control check as part of the production of cooked, cured meat products. It is also critical to the safety of many products to understand the interaction of different combinations of inhibitory factors on the germination and growth of *C. botulinum*. This is particularly relevant for cured meat products where nitrite, used as the inhibitory agent, is affected by other factors such as pH, presence of iron binding agents such as ethylene diamine tetraacetic acid (EDTA), the heat process and spore loading. In some cases it is possible that some so-called 'inhibitory' agents reduce antimicrobial activity in a product as a result of adverse interactions with other antimicrobials. This has been reported with isoascorbate where low levels may enhance inhibition of *C. botulinum* but high levels may cause nitrite depletion and thereby reduce the anti-botulinum effect (Tompkin *et al.*, 1978 and 1979a).

The heat process employed at the cooking stage will vary but, in the UK, a cook equivalent to 70°C for 2 minutes throughout the meat product must be ensured. This, of course, is primarily designed to destroy vegetative enteric pathogens. When cooking bulk meat, this heat process may take several hours to achieve an even cook throughout the product due to the size and density of the meat and the difficulty in achieving heat penetration to the centre. The outside of the meat is usually exposed to higher temperatures for longer periods of time than the centre although few of the heat process conditions will achieve significant destruction of clostridial spores. Temperatures of 70°C for 2 minutes will destroy vegetative cells of *C. botulinum* but will have little effect on the survival of the spores of non-proteolytic strains of *C. botulinum* and have no effect on the survival of spores of proteolytic strains. To achieve a significant reduction (6 log cfu/g) in even non-proteolytic strains of *C. botulinum* at 70°C would require all parts of the meat to be heated at 70°C for approximately 1670 minutes or 28 hours! (based on a calculation assuming 90°C for 10 minutes achieves a 6 log reduction and using a z value of 9°C).

The establishment and reliable application of appropriate heat processes is obviously important and validation studies are an essential pre-requisite for determining safe cooking processes for meat products. However, the usual heat processes applied to many meat products will not reduce levels of *C. botulinum* spores, if present, and growth controls need to be exerted during the further processing stages.

Some perishable, canned, cured meat products may be given heat processes greater than those for standard perishable, cooked meats but usually these do not exceed 90°C. To achieve safety with respect to non-proteolytic strains of *C. botulinum*, such products would need to be heated to 90°C for 10 minutes with preclusion of post-processing contamination, achieved as a result of cooking in cans.

The cooking process for chilled, cooked meat products, although designed to reduce the risk from vegetative microbial hazards, does affect the safety of the products in two other key ways. Firstly, heating spores to sub-lethal temperatures can stimulate the spores to germinate. This process is known as heat shock and this activation can occur by mild heat processes such as those achieved during pasteurisation. Secondly, due to the long cooling time after cooking, there is significant opportunity for both germination and growth of the organism at these warm temperatures. The large size of many bulk meats presents processors with significant challenges in trying to reduce the centre temperature of such products to below the growth range of proteolytic strains of *C. botulinum* in particular, and such times can exceed 4–6 hours. In general, cooked meat producers endeavour to reduce the temperature of cooked meat to <5°C within 6 hours but this is difficult to achieve for many producers of large size bulk meats.

Guidance provided in the United States Code of Federal Regulations allows large size bulk meats to be cooled from 48.8°C to 12.7°C (during which time growth is fastest) within 6 hours (United States Department of Agriculture, 1998) and most producers can achieve these targets. Likewise, the United States Food and Drug Administration requires that whole beef roasts and joints should be cooled from 60°C to 21°C within 2 hours and from 21°C to 5°C within 4 hours (Food and Drug Administration, 1999).

Similar advice has recently been published by Campden and Chorleywood Food Research Association (CCFRA) (Gaze *et al.*, 1998). A series of challenge test studies on cured and uncured bulk cooked meats assessed the danger presented by growth of surviving spore-forming clostridia (*C. perfringens*). Gaze *et al.* (1998) recommended cooling uncured meat products from cook temperatures to 50°C in 1 hour (maximum 2.5 hours), from 50°C to 12°C in 6 hours (maximum 6 hours) and from 12°C to 5°C in 1 hour (maximum 1.5 hours). It was considered that in cured meats the additional protective effect afforded by nitrite (ingoing > 100 ppm) and salt (> 2.5% aqueous phase) would allow all of these cooling parameters to be extended by 25%. However, it was also noted that cooling times would need to be significantly reduced if the initial spore loading of the meat mixture was high.

Some processors of large size bulk meats may find it difficult to achieve even these requirements, particularly when including further temperature elevation caused by subsequent roasting stages. In such circumstances it is important that processors demonstrate the safety of their cooling profiles by means of predictive mathematical modelling or challenge test studies, in experimental facilities using product formulations representative of their own.

The cooling stage of the process is critical to the safety of these products and all producers should have extensive information about cooling profiles built up from repeated cooking/cooling trials of their products. Cooling profiles should be determined using continuous temperature monitoring devices inserted into the part of the product likely to be subject to optimum bacterial growth temperatures for the longest time period; this is often the centre of the product after a full cook, but after a subsequent roasting process, this is likely to be closer to the surface of the meat (rather than the centre). Information should include a worst case situation using meat of the greatest thickness that has reached the highest internal cooking temperature and been exposed to the slowest cooling rate in normal operation. This should be conducted before any product is produced in a new manufacturing unit and should be part of the commissioning trials. It is also important to re-check cooling profiles at regular frequencies in normal production and, in particular, when new products are commissioned, if products are reformulated to be of a greater thickness or when new/different equipment is employed for cooling.

Juneja *et al.* (1997) conducted a series of experiments to determine the growth rate of a variety of bacterial pathogens during the cooling phase of cooked beef. Ground beef (pH 6.24) was inoculated with spores of three type B (non-proteolytic) strains and three type E strains of *C. botulinum*. The beef was heated to 60°C for 1 hour and then cooled from 54.4°C to 7.2°C over different time periods (6, 9, 12, 15, 18 and 21 hours). *C. botulinum* levels increased by only one order of magnitude over any of these time periods. It is assumed that this occurred over 21 hours, although this is not reported in the paper. Interestingly, when co-inoculated into the ground beef with *Bacillus cereus*, *C. botulinum* again increased over this temperature range by one order of magnitude over a 21-hour cooling period (Table 4.6). The authors concluded that this model food system indicated that cooling from 54.4°C to 7.2°C over a period of 21 hours would not pose a food safety hazard from growth of these pathogens. However, this needs to be perhaps more carefully considered as it is important to recognise that the strains used were non-proteolytic and it is possible that proteolytic strains may have grown more quickly under similar conditions. Of course, most chilling processes will be faster than this and as *C. perfringens* has been shown, under similar cooling conditions, to grow to in excess of 10^6/g if cooling times exceed 15 hours (Juneja *et al.*, 1994), cooling needs to be achieved in significantly less time than this to achieve overall safety.

Table 4.6 Growth of *C. botulinum* during the cooling phase of cooked ground beef, adapted from Juneja *et al.* (1997)

Time (h)	Temperature (°C)	Count* (log cfu/g)
0	54.4	3.55
4	34.6	3.53
6	27.8	3.73
8	22.5	4.03
10	18.3	3.98
12	15.1	4.15
14	12.5	4.08
16	10.5	4.24
18	8.9	4.58
21	7.2	4.35

* Variance Log 0.02–0.56 cfu/g
Six-strain mixture of non-proteolytic *C. botulinum* (three type B, three type E), together with *B. cereus*.
Beef previously autoclaved and then inoculated with heat shocked spores and heated to 60°C for 1 hour.

As the chilling process for the cooked meat is so important, it is essential that the blast chilling units used for such purposes are well maintained and capable of handling the volume of product introduced into them. It is sometimes the case that manufacturing units purchase and continue to use blast chillers and holding chillers which were designed for volumes of production significantly less than current volumes attained due to growth of the business. Under such circumstances the blast chiller will not be capable of adequately reducing the temperature of the products leading to extended periods of time where any surviving spores may germinate and grow. The limitations of the blast chiller need to be built into the hazard analysis to ensure that, as a critical process, they are not overloaded and, even if not overloaded, are maintained in a condition capable of achieving the objective of minimising the growth of *C. botulinum* and other surviving spore-formers.

Many cooked meat products are subject to a second heating stage during the process which is often referred to as roasting. Roasting usually takes place after the cooked meat product has been chilled and involves exposing the bulk meat to high temperatures in excess of 200°C for a short period of time in an oven or a flame tunnel. This produces a brown, roasted colour on the outside of the meat but, while the high oven temperature achieves an effective re-pasteurisation of the surface of the bulk meat, the roasting time is only sufficient to raise the internal temperature to 30–40°C. The time it takes to blast chill the product after roasting to temperatures outside the growth range of *C. botulinum* is again a critical control point in meats subject to a post-cook roasting. Indeed, when assessing the potential for the germination and growth of spore-bearing microorganisms like *C. botulinum*, it is important to take account of both the cooling stage after the primary cook and the heating and cooling stages of the meat in the roasting process.

Similarly, a practice often employed during the manufacture of comminuted meat mixes such as pâté is to add some of the ingredients such as the fat to the mix in a pre-warmed state. This may elevate the pre-mix temperature to within the growth range of *C. botulinum* and other pathogens present in the raw mix. While the pre-mix is usually processed quickly, it should be remembered that local areas of contamination may build up in 'dead spots', e.g. in the mixer or in pipes during a production day between full clean downs. Although the meat mix is then cooked, it is important to stress that such cooking is designed to deal only with limited levels of contamination and any significant build-up of contaminants can compromise the normally safe processing achieved by the pasteurisation temperatures applied. In such circumstances, either raw and pre-cooking product temperatures should be maintained below those allowing significant growth, e.g. < 10°C, or the frequency of cleaning should be increased.

Some cooked meat products are sliced and packed into retail pre-packs after cooking/roasting and this is often preceded by deep/super chilling to temperatures near 0°C for a short period to assist in slicing. Slicing is most often carried out using automated machinery. This involves placing the meat block onto a loading machine which is then inserted into the slicer.

The slicer cuts the meat into slices of defined thickness and the slices fall onto a conveyor belt that carries them to a packing machine where they drop into a plastic container which then has a plastic cover heat-sealed onto it.

The slicing operation has the potential to introduce environmental contaminants to the product from a number of sources including personnel, the slicing and conveying equipment and the general environment, i.e. ceilings, floors, walls and drains via aerosols. It is theoretically possible that spores of *C. botulinum* could be introduced to the product at this stage although this is not likely to be the primary route of the organism into the product, that being its introduction in raw materials and survival of the process. Nevertheless, it remains important to operate effective hygienic precautions which include high standards of operator hygiene practice and regular cleaning to minimise the extent of any contamination at this stage.

The temperature in slicing rooms is usually above 10°C and, as any delay may allow the bulk and sliced product temperature to increase, it is essential to minimise the time between removing the bulk meat from the chiller and re-chilling the sliced, packed meat. Understanding the effects of time delays on the elevation of product temperature is an important consideration that should be studied by any cooked meat processor. Time/temperature studies should be conducted using calibrated thermocouples to build in acceptable controls during the slicing stage, particularly in the event of any process breakdown and delay. Such systems may include taking the bulk meat back into chilled storage when a process delay exceeds a particular time period or indeed removing and discarding sliced, packed meat if left on the line for excessive periods. Criteria for this can only be set by individual processors as this must take account of other factors contributing to growth of the organism, including cooking, roasting and cooling times together with the formulation and shelf life of the product.

Products such as bulk pâté often have post-cooking garnishes added to them and, as garnishes often consist of herbs and spices, the potential for adding a high loading of spore-bearing bacteria must be taken into account. Inevitably, *C. botulinum* could be present in these garnishes and precautions should be taken to minimise the loading of spores on these materials, where possible, by washing prior to use or by the use of materials that have been subjected to a microbiological reduction process, e.g. superheated steam. It may also be useful to monitor the general spore loading of incoming batches of garnish to ensure levels do not exceed tolerable limits and, indeed, purchase specifications should reflect such limits.

Final product issues and control

The control of *C. botulinum* in chilled, cooked meat products is significantly affected by the primary processing conditions of cooking and chilling and the product formulation. However, one of the greatest potential problems associated with the safety of these products relates to the temperature control of the finished product after leaving the processing premises. The potential for any contaminating *C. botulinum* spores to germinate and grow is affected by three main factors; temperature, shelf life and the presence of any inhibitory factors in the product. Some cooked meat products have extremely long shelf lives in comparison to many other perishable, chilled products, often exceeding 30 days. Indeed, those chilled, canned, cured meats which require refrigeration after processing are often allocated lives of many months. This not only gives an increased potential for the non-proteolytic strains to grow, but also increases the chances that such products will be exposed to temperature abuse during their lives, either by the product being frequently removed from the refrigerator and replaced or as a result of the perceived stability in the eyes of the general public conferred by a product with a long shelf life. It is often assumed by those who do not fully understand the inherent hazards that products assigned greater shelf lives by the food industry are of greater stability (and safety!) than those with short shelf lives.

Under good conditions of temperature control (< 8°C), the primary hazards in relation to this product group are non-proteolytic strains of *C. botulinum* that are capable of growth under refrigeration conditions. However, if the product is exposed to temperature abuse involving temperatures in excess of 10°C, then it is possible that proteolytic strains of *C. botulinum* may also be of significance. In order to prevent the latter from occurring, it is essential that purchasers of cooked meat products intended to be stored under chilled conditions are made fully aware of the absolute requirement to keep them chilled. This is usually achieved by ensuring the pack of cooked meat is clearly labelled 'Keep refrigerated' and it is also common for some producers to state the maximum temperature of storage, e.g. 'Store below 5°C'. All too often however, such instruction is printed in very small characters, insufficient for many to see readily and so may go unnoticed. As it is such an important factor it is essential that such instruction is clearly legible and in bold print.

Most producers of cooked meats therefore assess the control of *C. botulinum* on the basis that the product will largely be kept under good refrigeration conditions and so only assess the hazard presented by non-proteolytic strains.

Growth of non-proteolytic strains of *C. botulinum* can be prevented if the pH of the product is 5.0 or less, if the aqueous salt content is 5% or greater or if the water activity is 0.97 or below. Therefore, if the product is formulated to achieve any of these factors individually, it could be considered safe with regard to the growth of this organism and an appropriate shelf life could be assigned accordingly. However, it is generally the case that cooked meat products have a pH between 5.8 and 7 and aqueous salt contents below 5%. Accordingly, the shelf life must be set taking account of the predicted growth rate of non-proteolytic strains of *C. botulinum* in the product of concern.

As cooked, uncured meat products do not contain nitrite, for some of these products the main factor contributing to preventing the growth of the *C. botulinum* is the aqueous salt content; growth may also be affected by the presence of acids such as lactic acid and possibly even emulsifying agents such as di- and polyphosphates, the latter exerting their effect by lowering the water activity of the product. In the presence of such variable factors, it is evident that judging the safe shelf life is not easy and it may be necessary to utilise predictive modelling approaches to assess the growth of the organism in products of differing formulations. This may be supplemented with challenge test studies to assess the actual growth in the product but the latter approaches are costly and provide no data for formulations other than the one challenged. The availability of models, such as Food MicroModel developed with funding from the UK Ministry of Agriculture, Fisheries and Food (MAFF) and those in the US developed by the United States Department of Agriculture (Pathogen Modelling Programme), have facilitated the prediction of shelf lives particularly with regard to psychrotrophic *C. botulinum*. However, in most cases such models predict fairly rapid growth of psychrotrophic *C. botulinum* in uncured meat products, e.g. in a product containing 1.8% aqueous salt and with a pH of 6.2; at 5°C, *C. botulinum* would be predicted to grow from an initial contamination level of 1 cfu/g to 10^3 cfu/g within 17 days (Food MicroModel, 1999). This, of course, does give those in the food industry cause for significant concern as many products with this typical formulation exist with shelf lives well in excess of this. This is not actually a deficiency of the specific model itself, but more the lack of the model's ability to include all factors likely to contribute to restricting the growth of the hazard.

It is probable that as many cooked, uncured meat products have chilled shelf lives in excess of 17 days and that none have been implicated in outbreaks of botulism due to growth of non-proteolytic *C. botulinum*, other factors may exist which contribute to their safety. Such factors probably include acidity and polyphosphates, not currently modelled in many predictive modelling systems. Indeed, it may also be the case that growth of some microbial pathogens in structured foods like cooked meats may be restricted by the microstructure of the food itself, as immobilisation of microbial pathogens in structured foods is known to affect the growth rate (Robins *et al.*, 1994).

However, that *C. botulinum* can grow in cooked meats is not in question; the big question is, how quickly can they produce toxin? Research carried out at Campden and Chorleywood Food Research Association in the UK, demonstrated the potential for non-proteolytic types B and E of *C. botulinum* to grow and produce toxin in previously autoclaved, cooked chicken homogenates. Spores were inoculated into the homogenates and then subjected to the normal heat process given to these products, a cook equivalent to 2 minutes at 70°C. Samples were then stored at the test temperatures and examined at regular intervals for levels of the organism and toxin production. In a series of experiments (Brown and Gaze, 1990 and Brown *et al.*, 1991) the potential for both types to grow and produce toxin was evident, the time to toxin production being related to the temperature of storage (Tables 4.7 and 4.8). At 5°C, toxin was detected in only one sample after storage for 12 weeks. However, toxin was detected in a sample inoculated with type B at 8°C after only 6 weeks storage and after 8 weeks for type E. Storage at 15°C allowed toxin production by both types B and E after only 1 week, demonstrating the significant hazard associated with temperature abuse.

It is most interesting to note that toxin production was evident in some cases in the absence of demonstrable increases in the population of *C. botulinum*. This may have been due to accumulated toxin produced by the low population or due to difficulties in the enumeration of the organism in the product matrix. It is however recognised that specific detection and enumeration of *C. botulinum* in real food matrices is not entirely reliable using cultural techniques. However, such data should sound warnings to those using growth predictions for the assessment of the risk with regard to this organism as, even without significant apparent growth, toxin was detected.

Other workers have also studied the growth of non-proteolytic strains of *C. botulinum* in cooked, uncured meats. Meng and Genigeorgis (1994)

Table 4.7 Growth of *C. botulinum* types B and E in chicken homogenate, adapted from Brown and Gaze (1990)

Time (weeks)	*C. botulinum* type B (log count/g – mean of three replicates)			*C. botulinum* type E (log count/g – mean of three replicates)		
	Temperature			Temperature		
	3°C	5°C	8°C	3°C	5°C	8°C
0	ND	ND	ND	ND	ND	ND
2	0.48	0.9	0.6	3.4	3.43	3.44
4	0.3	0.48	0.3	3.29	3.38	3.37
6	0.3	0.3	5.12*	3.27	4.22	2.85
8	2.77	2.67	4.97*	3.47	3.36	2.86*
12	0.6	1.18	ND	3.26	3.38	2.94* †

* Toxin detected.
† Toxin also detected at week 10 (log count 3.14/g).
ND = not determined.
Inoculum: c. 100 spores/g (type B) and c. 1000 spores/g (type E).
Products were inoculated, subjected to a heat process equivalent to 70°C for 2 minutes and stored at 3°C, 5°C or 8°C.
Levels after heating: type B 3.8×10^1/g, type E 3.3×10^3/g.

demonstrated growth and toxin production by a mixture of type B and E strains in cooked, uncured chicken and beef. Inoculum levels of 10^4 spores (inoculated into 3 g of meat) produced toxin at 4°C, 8°C and 12°C after 90 days, 8 days and 4 days, respectively in beef, and 90 days, 16 days and 12 days, respectively in chicken (Table 4.9). In the same study they investigated the effect of sodium lactate on toxin production and found that levels of 2.4% in beef and 1.8% in chicken significantly delayed toxin formation (Table 4.9).

Meng and Genigeorgis (1993) also reported on studies of the growth of non-proteolytic *C. botulinum* types B and E in cooked turkey at a range of temperatures with varying levels of sodium chloride and inoculum sizes. They demonstrated that varying inoculum sizes had a noticeable effect on time to toxin formation, which in many cases was more pronounced than the effect of varying salt concentration (Table 4.10).

It is therefore considered prudent that cooked, chilled meats, stored under good conditions of refrigeration and containing no added preservatives such as nitrite, in which the salt concentration is less than 3.5%, should be given shelf lives of no greater than 15 days due to the potential for growth of non-proteolytic strains of *C. botulinum*. Products may be

Table 4.8 Growth of *C. botulinum* types B and E in chicken homogenate, adapted from Brown *et al.* (1991)

Time (weeks)	*C. botulinum* type B (log count/g – mean of three replicates)						*C. botulinum* type E (log count/g – mean of three replicates)					
	Temperature						Temperature					
	5°C		8°C		15°C		5°C		8°C		15°C	
	Type B1	Type B2	Type B1	Type B2	Type B1	Type B2	Type E1	Type E2	Type E1	Type E2	Type E1	Type E2
0	NT	NT	NT	NT	NT	NT	NT	NT	NT	NT	NT	NT
1	NT	NT	NT	NT	5.6*	2.53*	NT	NT	NT	NT	5.14	5.2*‡
2	NT	NT	NT	NT	6.7*	3.4*	NT	NT	NT	NT	5.94*	5.74*
4	NT	NT	1.72	0.3	NT	NT	NT	NT	1.53	2.86*†	NT	NT
6	0.7	0.3	2.43	0.3	NT	NT	1	2.27	2.38	2.59	NT	NT
8	NT	0	1.15	0.3*	NT	NT	NT	NT	2.73	4.07	NT	NT
12	0.6	0.3*	0.78	0.3*	NT	NT	2.27	3.32	2.5	5.49*	NT	NT

*Toxin detected.
† Levels after week 3 were log 2.68/g (no toxin detected).
‡ Levels after 0.7 week were log 3.3/g (no toxin detected).
NT = not tested.
Inoculum: 1000 spores/product.
Products were inoculated, subjected to a heat process equivalent to 70°C for 2 minutes and stored at 5°C, 8°C or 15°C.
Levels after heating: B1 3.4×10^1/g, B2 5/g, E1 2.0×10^2/g, E2 2.1×10^2/g.

Table 4.9   Toxin production by *C. botulinum* types B and E in cooked beef and chicken (10^4 spores/3 g) at $4°C$, $8°C$ and $12°C$ and the effect of added sodium lactate, adapted from Meng and Genigeorgis (1994)

	Inhibitory factors			Time to toxin production (days)		
				Temperature		
	Sodium lactate (%)	Brine concentration (%)	pH	$4°C$	$8°C$	$12°C$
Beef	0	0.96	5.8	90	8	4
	2.4	0.97	5.8	>90	90	>40
	4.8	1	5.8	>90	>90	>40
Chicken	0	0.52	5.9	90	16	12
	1.8	0.52	5.9	>90	60	>40
	3.6	0.54	5.9	>90	>90	>40

capable of achieving longer shelf lives but, before allocating longer life, it is recommended that such products should be challenge tested to determine their capacity to inhibit the growth and toxin production by *C. botulinum*. Alternatively, the relevant inhibitory factors should be entered into predictive mathematical models to demonstrate that such a shelf life would be inherently safe under chilled storage conditions.

Cooked, cured meats, i.e. those containing sodium nitrite, are more inhibitory to the growth of *C. botulinum*. Much of the research on the

Table 4.10   Time to toxigenesis of *C. botulinum* types B and E in cooked turkey roll at varying temperature, salt and inoculum levels, adapted from Meng and Genigeorgis (1993)

	Time to toxin production (days)		
	Salt (aqueous concentration %)		
	0.5	1.47	2.2
Temperature: 8°C			
Inoculum: 10^4 spores/g	7	7	8
Inoculum: 10^2 spores/g	8	14	16
Temperature: 12°C			
Inoculum: 10^4 spores/g	3.25	5	7
Inoculum: 10^2 spores/g	5	7	9

Product formulation: <0.5% polyphosphate, pH 6.2-6.3.

anti-botulinum properties of nitrite has been carried out using proteolytic strains. Work with non-proteolytic types has been conducted but primarily in investigating their growth under temperature abuse situations, usually involving higher (> 10°C) temperature studies. It is evident from this work however that *C. botulinum* is significantly inhibited by this preservative and although the research has been extensive in this area, the absolute mechanism by which inhibition is effected is not fully understood. Nitrite is believed to have most effect in delaying outgrowth of the organism and restricting subsequent growth and only limited effect in delaying germination of the spore itself. It is variously suggested that its inhibitory activity is associated with interference with iron availability to the cell, disruption of membrane function, oxidation of cellular biochemicals and interference in energy transfer mechanisms.

Nitrite is known to be depleted rapidly at low pH or at high ambient temperatures, although it may form additional anti-botulinum compounds in meat at cooking temperatures. Therefore, it is still debated whether it is better to monitor ingoing or residual nitrite in cooked meat products. In the authors' opinion, it is best to control the process based on ingoing nitrite between 75 ppm and 150 ppm while ensuring some is present (> 10 ppm) immediately after processing. The presence of iron can decrease the efficacy of nitrite and chemicals that bind iron such as iso-ascorbate and EDTA are reported to increase the efficacy of nitrite (Tompkin *et al.*, 1979b). Care must therefore be exercised when using those products naturally rich in iron such as liver and heart. Also, it is important to recognise that high ascorbate/isoascorbate levels (> 400 ppm) may actually reduce the inhibitory effect of nitrite by depletion (Tompkin *et al.*, 1978 and 1979a). Although nitrite may be more rapidly depleted at lower pH values, it is also reported to be more active at lower pH which supports the theory that undissociated nitrous acid may play a significant role in inhibition of *C. botulinum*.

Nitrite does not completely prevent the growth of *C. botulinum* but merely extends the lag time and reduces the rate of growth.

To combat depletion, some processors add nitrate in the product formulation as well as nitrite, as the former can provide a 'reservoir' of nitrite, the nitrate becoming reduced during the shelf life due to the activity of nitrate-reducing bacteria in cooked sliced meat. However, this mechanism is not a reliable one as it is dependent on the presence of contaminating microorganisms in the final product. Reliance on such factors can be undermined by improvements in processing hygiene which may reduce contamination with nitrate-reducing bacteria; therefore use of nitrate should be restricted to meats where such bacteria will invariably be present, such as raw, cured meats like bacon and salami.

Little work has been published on the growth of non-proteolytic strains of *C. botulinum* in cured meat products although Pivnick and Bird (1965) conducted some useful research on the growth of type E in chilled cooked meats under abuse conditions. Their research clearly demonstrated the effect of differing spore loads on the potential for growth and toxin formation by *C. botulinum* (Table 4.11) with levels of 10 000/g of meat being far more likely to result in toxic hams after shorter periods of incubation at 10°C and 15°C.

Table 4.11   Effect of spore load and temperature on toxin formation by *C. botulinum* type E in jellied ox tongue, adapted from Pivnick and Bird (1965)

Temperature (°C)	Spore load (cfu/g)	Number of toxic packs				
		Time (weeks)				
		0.5	1	2	4	8
15	10	—	1/5	4/5	—	—
	100	—	0/5	5/5	—	—
	10 000	3/5	2/5	5/5	—	—
10	10	—	—	—	0/5	0/5
	100	—	—	0/5	1/5	2/5
	10 000	—	—	2/5	3/5	5/5
5	10	—	—	—	0/5	0/5
	100	—	—	—	0/5	0/5
	10 000	—	—	0/5	0/5	0/5

Product formulation: aqueous salt 2.4%, nitrite 55.5 +/- 39 ppm, pH 6-6.4.
Vacuum packed in oxygen-impermeable packaging.
No positives in uninoculated control samples.

They also clearly demonstrated the significant potential for toxin formation in cured meats under conditions of temperature abuse. At temperatures of 10°C and above, packs of meat became toxic, with the speed of toxin production being progressively faster at higher temperatures (Table 4.12). It is interesting to note that while their work showed that *C. botulinum* type E could grow and produce toxin in many commercial meat formulations (Table 4.13), they did not detect any toxin formation at temperatures of 5°C, although the work extended to periods of only eight weeks (Table 4.12).

Clostridium botulinum

Table 4.12 Effect of temperature on toxin production by *C. botulinum* type E (5 strain mixture) in cooked sliced ham, adapted from Pivnick and Bird (1965)

Temperature (°C)	Toxin formation (number of packs)				
	Time (weeks)				
	0.5	1	2	4	8
5	—	—	—	—	0/10
10	—	—	—	0/10	0/10
15	—	0/10	0/10	0/10	1/10
20	0/10	0/10	1/10	3/10	—
25	0/10	1/10	4/10	—	—
30	2/10	2/10	—	—	—

Inoculum: 10 000 cfu/g inoculated between pack slices. Uninoculated controls showed no signs of toxin formation.
Vacuum packed in oxygen-impermeable packing.
Formulation (estimated from report): aqueous salt 3.1%, sodium nitrite 97 +/- 31 ppm, pH 5.8–6.4.

Lücke *et al.* (1981) conducted some challenge test work with spores of non-proteolytic *C. botulinum* inoculated into liver sausage (83 ppm added sodium nitrite, a_w 0.98, pH 6.2) and bologna type sausage (83 ppm added sodium nitrite, a_w 0.985, pH 6.1) prior to cooking to temperatures of 76°C for 4 minutes and storage for 110 days at temperatures between 5°C and 15°C. They found that while toxin was detected after 12, 18 and 28 days at 15°C, 10°C and 8°C respectively, the level of nitrite offered variable protection against toxin formation.

The work of Pivnick and Bird (1965) also demonstrated an extremely important principle of major significance to the food industry in relation to the growth potential of *C. botulinum* in food products, although not related to inhibition by nitrite. They clearly showed that growth and

Table 4.13 Toxin production by *C. botulinum* type E in commercial cooked sliced ham after seven days storage at 30°C, adapted from Pivnick and Bird (1965)

Manufacturer	Aqueous salt (%)	Sodium nitrite (ppm)	Toxin production (number of packs)
a	3.1	97 +/- 31	1/7
b	4.3	87 +/- 72	1/7
c	3.4	47 +/- 38	1/7
d	3.4	47 +/- 24	2/7
e	3.5	130 +/- 69	0/7

Inoculum: 10 000 spores/g.

toxigenesis were not affected by the types of packaging or gaseous atmosphere in the pack of cooked meat and that toxin could be produced in oxygen-permeable packed meat as quickly as it could in impermeable vacuum-packed or anaerobically stored meat. The factors governing growth are thus more closely related to the conditions in the micro-environment in the product rather than the gaseous conditions in the surrounding atmosphere.

In a cooked meat the organism occurs on or in the meat, which at the subsurface level it has little oxygen and is generally in an anaerobic environment with very low redox potential, even in air-packaged cooked meat. Therefore, even if stored under oxygen-containing atmospheres, meat and many other foods have the potential to support the growth of *C. botulinum*.

It is clear therefore that while it is possible to indicate formulations that will inhibit the growth of non-proteolytic *C. botulinum* under effective chilled conditions, i.e. aqueous salt 5% or more, a_w 0.97 or less, etc., when formulating products with levels less stringent than these, the combination of factors controlling growth requires more detailed assessment. Although primarily based on work assessing the effects of anti-botulinum factors on proteolytic strains under ambient conditions it is clear that certain factors can make cooked meat products safer with respect to the hazard of the growth and toxin production by non-proteolytic *C. botulinum* during the shelf life of the product.

Safer	Less safe
Temperature $< 5°C$	Temperature $> 10°C$
pH < 6	pH 6 or greater
Nitrite (75–150 ppm ingoing)	No nitrite
Isoascorbate (> 50–< 200 ppm)	High levels of isoascorbate (> 400 ppm)
Presence of polyphosphates or citrates	No polyphosphates or citrates present
Presence of iron chelators (EDTA)	Presence of iron
Salt ($> 3\%$ aqueous)	No salt or reduced salt ($< 1\%$)

Clearly, the use of combinations of these factors will result in greater inhibition of growth than the use of individual factors alone.

It is recommended that cooked meat products, to which nitrite is added as a preservative (75–150 ppm) and where no single controlling factor is present which prevents growth, e.g. >5% aqueous salt, should be allocated chilled shelf lives of no greater than 30 days, unless specifically shown to be capable of preventing the growth of non-proteolytic strains of *C. botulinum*. Clearly, products may be allocated longer shelf lives, but it is the opinion of the authors that such lives should be justified by the use of challenge testing or predictive mathematical models which clearly show the safety of the product given the intended formulation and shelf life.

In order to maintain the safety record of these products with respect to *C. botulinum*, a full and structured hazard analysis of the specific manufacturing process, product formulation and shelf life/product use conditions is essential together with the implementation and consistent maintenance of all critical controls.

AMBIENT-STORED, MINIMALLY PROCESSED VEGETABLES IN OIL

The decision to purchase a food has much to do with the visual appearance of the product to the purchaser. Product packaging has become much more sophisticated to try to attract the eye of the buyer and many products such as fresh fruit and vegetables are entirely dependent on being seen in a near perfect condition before they will be bought. In recent years there has been a significant increase in the manufacture and sale of vegetables in oil. These products are packaged in transparent bottles and consist of a mixture of herbs, spices and vegetables which are chosen to give a flavour to the oil in which they are stored but often achieve sales because of the appealing visual characteristics that they portray on the shelf. Packed with red peppers, fresh garlic and green spring onions, these products have a fresh and aesthetically appealing attraction to them that has resulted in increasing sales. In addition, as they appear to the untrained eye to be very easy to make, it is common for these products to be manufactured in small farmhouse type operations to unsophisticated recipes. Indeed, because of their visual appeal, it is also common to see these products being sold not only by food retailers but also by home furnishing shops who see these products as a way of brightening up the kitchen by putting them on display and adding to the 'design features' of the house. Vegetables in oil are normally designed to impart flavour to the oil which may be used for dressing salad or for frying other foods. Although the oil is always consumed, it is also common for the vegetable components to be consumed as well.

Description of process

Vegetable in oil products can be made from virtually any combination of herbs, spices and vegetables (Figure 4.2). They frequently include garlic and chilli peppers which may be used whole or sliced but may also be made with artichokes, aubergine (eggplant), courgette, spring onion, mushrooms, and so on. The key criterion is usually whether it will provide a good visual appearance in the bottle on display and/or whether it will impart a pleasant flavour to the oil. The vegetable is usually supplemented with a selection of herbs such as lemongrass, mint leaves, bay leaves and spices such as peppercorns.

Processing of the product varies considerably with many vegetables such as garlic and spring onions being peeled and washed to remove visible soil debris and then either placed whole into the oil or sliced and placed into oil. Products such as whole chilli peppers are added to oil either fresh or after partial sun drying. In some cases, the vegetable may be soaked in acid

Process Stage	Consideration
Raw vegetable growing	Agricultural practice
↓	
Harvesting	Hygiene
↓	
Transport and storage	Hygiene
↓	
Peeling/trimming/preparation	Hygiene
↓	
Washing (if applicable)	Chlorination
or	
Drying (if applicable)	Time
	Temperature
or	Final water activity
Cooking (if applicable)	Time
	Temperature
or	
Salting/acidifying (if applicable)	Level of preservative
	Distribution of preservative
↓	
Slicing/further processing (if applicable)	Hygiene
↓	
Addition to oil	Hygiene
↓	
Retail sale	Shelf life and storage conditions
↓	
Customer	Storage conditions

Figure 4.2 Process flow diagram and technical considerations for a typical vegetable in oil product.

or a brine solution prior to placing in oil. The vegetable component is often cooked prior to placing it whole or sliced into the oil. For example, garlic is often baked in foil as a whole head of garlic before the individual cloves are separated and placed with or without removing the outer skin into the oil. Artichokes and aubergines are usually baked and mushrooms may be boiled before placing in oil. What is usually common to all these processes is that the cooking stage, if present, does not take place in the oil and the vegetable is handled or prepared by peeling or slicing prior to being placed in the oil.

Herbs and spices are frequently added to the oil either fresh, as would be the case for mint and lemongrass, or in a dried form such as peppercorns. Mushrooms may also be dried or occasionally may be brined prior to adding to the oil. The oil used for these products is usually a vegetable oil or, more frequently, olive oil.

The products are stored in transparent glass containers and sold under ambient conditions with shelf lives exceeding 12 months and often up to two years.

Raw material issues and control

Vegetables, herbs and spices are unquestionably going to be contaminated with spores of *C. botulinum*. This is due to the nature of the conditions in which these materials are grown and harvested and the fact that little reduction in contamination will generally occur during further processing. Vegetables are grown in soil which is known to be frequently contaminated with spores of *C. botulinum* (Table 4.14) and it is inevitable that during growing and harvesting some of the soil and associated microflora will be transferred to the plant. Although washing, if employed, may reduce the level of microbial loading, it is not possible to remove all microorganisms and as spores will be a major component of the microbial populations then *C. botulinum* will, on occasion, be present. Therefore, the relevant question in relation to raw material contamination with *C. botulinum* is related to the levels and frequency of its presence rather than contamination itself. Notermans (1993) reviewed the incidence of *C. botulinum* in vegetables and reported contamination ranging from 0% to over 40% with levels in mushrooms as high as 41 spores per 100 g.

Table 4.14 Incidence of *C. botulinum* in vegetables, adapted from Notermans (1993)

Product	Incidence (%)	Types	Levels
Vegetables (USSR)	13/30 (43.3)	A, B	NR
Spinach (USA)	6/50 (12)	A, B	NR
Green beans (USA)	0/50 (0)	—	—
Onions (USA)	5/75 (6.7)	A	NR
Carrots (Hungary)	0/18 (0)	—	—
Mushrooms (Canada)	NR	B	41/100 g

NR = not recorded.

The types of *C. botulinum* present relate to the incidence in individual soil types and countries (Table 1.5) but types A and B are most frequently associated with soil and vegetation and therefore are likely to predominate in these raw materials.

In addition to the vegetables, many vegetable in oil products contain herbs and spices sourced from throughout the world. Although these are likely to be subject to similar contamination routes as vegetables, it is often the case that many herbs and spices are imported from poorly developed countries where the standards of agricultural and hygienic practice may be significantly less well advanced than those in developed countries. In addition, many herbs and spices are dried in the sun in the producing countries with little protection from contamination. As a result of the dry, dusty conditions, extensive contamination can occur with spore-forming microorganisms carried in soil dust to materials being dried. Such material can be heavily laden with spore-bearing microorganisms and, while they present little hazard to the commodity in its dry form, such material may be the source of high levels of contamination for the finished oil based products. Although surveys of the incidence of *C. botulinum* in herbs and spices are limited, it is clear that with usual spore loads of 10^4/g or more, comprising both *Bacillus* species and *Clostridium* species, a few will inevitably be *C. botulinum*.

It is not possible to significantly reduce the level of contaminating spore-forming organisms in primary agricultural vegetables or herbs and spices although, wherever possible, good standards of agricultural practice will help minimise the extent of contamination occurring. Therefore, subsequent processing must take account of such hazards introduced by the raw materials and ensure that the subsequent product formulation or processing either destroys the organism or prevents the ability of surviving contaminants to germinate and grow.

Process issues and control

Processing of the vegetable raw materials for these products ranges from none at all to washing/peeling to the introduction of destruction and preservation strategies.

Herbs and spices may be added to oil fresh or after drying. Although there are numerous reports of natural microbiologically inhibitory compounds in a variety of spices it is clear that such inhibition is not well understood and may be subject to variation depending on factors such as processing, packaging and other components/conditions in the final product. Fresh cut herbs, in particular, and some spices release nutrient components from cut surfaces and may actually support the growth of *C. botulinum*. However, other spices are derived from bark, roots or from seed pods where the moisture content and consequently the water activity is very low and thus less likely to support the growth of *C. botulinum*. Without definitive data on specific growth-controlling components in herbs and spices or without conducting challenge testing of the finished product, it is not possible or sensible to preclude the possibility of growth of *C. botulinum* and toxin production when herbs and spices are used in these oil based products.

Fresh herbs and spices are usually washed to reduce soil and insect contamination on the surface, although on vegetables such washing has been found to reduce microbial loading by no more than one order of magnitude (Adams *et al.*, 1989). In addition, washing adds moisture to the material, increasing the risk of growth of organisms associated with the material in the final product in oil.

The safest approach to ensuring control of the growth of *C. botulinum* is to ensure factors are present in the raw material or finished product that are demonstrably capable of preventing growth. For example, a material may have a water activity at its surface below the limit of growth for proteolytic *C. botulinum* (0.94 or below) or a pH value close to the minimum for growth of the organism. By understanding the basic factors inherent in the raw material itself it may be possible to select raw materials capable of preventing growth of the organism without changes in product formulation, while avoiding those raw materials with inherently greater risk. It should be remembered that many of these raw materials may be subject to significant variation in moisture and pH as a result of climatic differences in growing and harvesting conditions and, if these controlling factors are to be relied upon in isolation, it is essential to monitor closely the pH or moisture/water activity of incoming batches to ensure they comply with specified limits. One of the best approaches for ensuring safety is to use naturally dry of artificially dried herbs and spices. Such material is inherently stable due to the very low moisture levels achieved during drying (c. 10–13%) which reduces water activity values to below 0.8, well below the minimum to prevent germination and growth of *C. botulinum*.

Of significantly greater concern in relation to the hazard of *C. botulinum* are the vegetable components used for these products. Many of the vegetables used for these products have very high water activity and the pH of the material is often well within the growth range of *C. botulinum*.

Processing of vegetables can range from none at all, to washing in water or to fully baking and slicing. Products receiving little or no processing are generally those vegetables, such as spring onions, which cannot tolerate any form of cooking if they are to maintain their appearance in the bottle. Materials such as these may be heavily laden with soil in the root system and also contamination can occur within the outer layers of the leaves due to airborne contamination from dust and water. Reducing contamination in such materials can only be achieved by cutting the roots off and peeling the external layers of the onion away and then washing the remaining material. However, such processing merely reduces and certainly does not eliminate the potential for *C. botulinum* to be present and placing such products into oil provides no assurance that any contaminating *C. botulinum* will not germinate and grow.

Although little data are available on the potential for growth of *C. botulinum* in spring onions, a number of studies have been conducted on the hazard from *C. botulinum* in onions. These show that growth and toxin production can occur when the material is stored at ambient temperature under anaerobic conditions (Austin *et al.*, 1998). The potential for growth in these materials will be dependent on the level of inhibitory factors present such as pH, water activity and presence of naturally inhibitory substances. In the majority of cases none of these factors are present at levels capable of preventing the growth of *C. botulinum* and toxin production, given that these products will be stored under otherwise optimum conditions, i.e. at ambient temperatures under anaerobic conditions in oil. Vegetables receiving no further processing prior to being placed in oil are therefore a direct botulism hazard and it is surprising that governments do not act to enhance the safety of such products, the notable exceptions being the US and Canada.

A number of vegetables are subject to some form of cooking process prior to further processing. In most cases, the cooking process is designed to make the raw material palatable prior to adding to the oil rather than to destroy *C. botulinum*. Most of the cooking processes are therefore quite mild and achieve temperatures to soften the vegetable material and destroy vegetative spoilage microorganisms like yeasts. For example, garlic may be cooked as whole bulbs in foil at oven temperatures of 200°C for up to one hour. Likewise, this process may also be applied to aubergines or artichokes. Although the oven temperature is very high, the internal product temperature rarely exceeds 100°C and the level of spore destruction will vary from the surface of the product to the centre. Some spore destruction may occur on the skin which is where most contamination is likely to occur but such processing will rarely eliminate the organism completely and heat processing to achieve sterility would inherently destroy the visual appearance of the product.

Heat processes of the nature applied to garlic are often employed to destroy contaminating yeasts and other anaerobic/facultatively anaerobic bacteria capable of growth in and spoilage of the product in the oil. While this does achieve the objective of destroying vegetative microbial contaminants, it unfortunately leaves the greater hazard of surviving spore-forming bacteria such as *C. botulinum.* Clearly, if sterilisation processes were employed to treat the raw vegetable material then, providing this was employed effectively, it could render the raw material and finished product safe when subsequently aseptically added to the oil. This however is unrealistic and cooking processes used for these vegetables do not achieve sterility. Also, contamination is likely to occur during the subsequent handling of the cooked product as it is added to the oil. Post-heat-process contamination can be further exacerbated if the vegetable is subsequently sliced or otherwise processed.

Baked products such as garlic are cooled and then individual cloves are broken off the bulb and either placed whole into the oil or peeled and sliced and then added to the oil as slices. Such handling introduces opportunities for re-contamination and, importantly, slicing also releases nutrients from the product which become available for microbial growth. In a situation where the material has had most microbial competition removed by mild cooking, any surviving spores are likely to face very little competition from growth of other microorganisms. As such, the risk from these products is magnified as the contaminating pathogen can grow unrestricted by competitive growth. In addition, vegetative contaminants, if present, could spoil the product before toxin production by *C. botulinum* could occur and in destroying them in a previous heat process, a possible indicator of product instability is lost.

The only effective and reliable way of making both the vegetable and herb/spice component of these products safe is to ensure that they are formulated to prevent the germination and growth of *C. botulinum.* Many products of this nature are formulated in this way by one of two key strategies. Firstly they may have the water activity reduced by drying or by the addition of humectants or they may be acidified to reduce the pH.

Many vegetables are dried prior to adding to oil and this is an extremely effective way of ensuring safety and preventing spoilage. Products such as garlic can be sliced and dried and chilli peppers can be stabilised with mild sun or oven drying without losing visual characteristics. Products must be dried to water activity levels of 0.94 or below and this is readily achievable for many vegetables. Some other components found in these products, like tomatoes, are often dried to reduce water activity. Sun dried tomatoes, of course, also benefit from having a low pH which provides an additional safety barrier to the growth of *C. botulinum*.

Although affecting product flavour slightly, it is quite possible to achieve complete safety by acidifying the vegetable prior to addition to oil. Garlic, aubergine and many others can be acidified to a pH of 4.6 or less to preclude the growth of *C. botulinum*. Under the conditions of oil storage, growth of moulds and concomitant elevation of local pH is precluded, although potential spoilage and/or pH elevation by other spore-forming anaerobes may be of concern (see section 'Pasteurised, acidified, ambient-stored fruit and vegetables' in this chapter). To ensure safety, it is essential that the acidification process achieves equilibration throughout the vegetable and it may be necessary to leave the vegetable in acid for extended periods to allow transfer of the acid and reduction of pH to the required level throughout the material.

The type of acid used is, of course, also important in relation to the control of *C. botulinum*. Acidification using acetic acid is usually the chosen option but some vegetables may discolour in the presence of this acid and other acids may need to be used. Therefore, it is important to recognise that not all acids are equally effective in inhibiting the growth of *C. botulinum*. Under such circumstances it may be necessary both to challenge test the formulation and control the organism using a lower pH, e.g. < 4.0. As it is only the vegetable component that is acidified, there is little effect on the flavour of the oil and, therefore, this is an accepted approach for ensuring the safety of these products.

Another option often employed in the safe manufacture of these products is the use of salt. Some raw materials such as mushrooms can be made safe by elevating the salt content to a sufficiently high level to preclude the growth of *C. botulinum*. Such levels would need to achieve 10% or greater aqueous salt and would normally be achieved by soaking in a brine solution for extended periods to achieve equilibration before placing the food in oil.

Whichever method is adopted to control the growth of *C. botulinum* in these products, it is clear that the factor(s) used must be subject to rigorous process control to ensure the controlling factor is achieved at the appropriate level throughout the raw material. This should be allied to a high level of process and material inspection using analyses for pH, aqueous salt or water activity.

Final product issues and control

The potential for growth and toxin formation by *C. botulinum* in vegetable products stored in oil is entirely dependent on the processing conditions employed to control the organism and the formulation of the finished product. Many products currently displayed on the shelves of some retail stores are not formulated to ensure complete safety and it is surprising that more outbreaks have not occurred from these products. However, there are so many variable factors involved in achieving safety that it may be that the unfortunate circumstances required for an outbreak occur only rarely. For example, factors such as the presence of *C. botulinum* in product components, the presence of nutrients capable of supporting growth and toxin production, the location of the organism in relation to these nutrients, the lack of naturally inhibitory factors, the lack of process inhibitory factors such as low pH, high salt or low moisture, the use of the product over an extended period and consumption of the toxic component without further heating all have to come together to make the potential hazard become a real outbreak or incident of botulism. In any case, it is incumbent on the food industry in this current age of hazard analysis approaches to develop, produce and sell products to consumers that are designed to be safe based on comprehensive processing and product control rather than on luck.

There is no doubt that some vegetable in oil products can and do support the growth of *C. botulinum*, although it is suggested that of greatest concern are the proteolytic strains as these are more adapted to the conditions prevailing in the processing and preservation of vegetables (Notermans, 1993).

Following the outbreak of botulism associated with chopped garlic in oil, Solomon and Kautter (1988) studied the potential for *C. botulinum* to grow and produce toxin in sliced garlic stored in soybean oil. The product manufactured from dried garlic slices, rehydrated and added to oil, was inoculated with spores of type A and type B *C. botulinum* at levels of 1 spore/g and 5 spores/g (Table 4.15). At 35°C, toxin was produced by type A within 15 days and by type B within 20 days at an inoculation level of 1 spore/g. At the higher inoculation level of 5 spores/g of garlic, toxin was produced in several samples after 15 days by both type A and B when stored at room temperature.

Table 4.15 Growth of *C. botulinum* types A and B in garlic in oil, adapted from Solomon and Kautter (1988)

Type	Temperature	Time to toxin production (days) (number of samples toxic/number examined)				
		15	30	45	60	75
A*	RT	13/25	17/25	17/28	16/24	16/24
B†	RT	2/20	7/20	8/20	6/18	7/18

* Average of eight strains, each inoculated at 5 spores/g.
† Average of seven strains each inoculated at 5 spores/g.
RT = room temperature.

In a further study of a variety of bottled vegetable products stored in oil, Solomon *et al.* (1991) found that no products became toxic after four months' storage at ambient temperatures (23°C) when inoculated with a 5 strain mixture of *C. botulinum* type A at a level of 50 spores/g or /ml. Products included oil extract of garlic, black beans in oil, chilli-garlic in oil, chopped shallots in oil, walnuts in oil, sun-dried tomatoes in oil, dried tomatoes in olive oil, garlic in water, dried tomatoes in sunflower oil and pesto sauce. The pH of all the products was 6.0 with the exception of the chopped shallots which had a pH of 4.7 and the tomatoes in oil (pH 4.0–4.5). In assessing the significance of these findings it is important to recognise that the previous processing of these products is not reported and clearly the potential to inhibit the production of toxin may have had much to do with the salt content and/or water activity of the products together with any other inhibitory factors not reported. Certainly, the low pH of the shallots and the tomatoes could have precluded toxin production by *C. botulinum* and perhaps the presence of salt or low moisture may have provided some protection for the other products.

As many of these products have cooked vegetables added to the oil, the work of Carlin and Peck (1995) gives significant weight to the potential dangers associated with these types of products. They inoculated 28 cooked, puréed vegetables with a 5 strain mixture of proteolytic *C. botulinum* types A and B and also investigated the growth of non-proteolytic strains of types B, E and F (11 strains). All of the products were inoculated with c. 10^3 spores/g and incubated at 30°C for 60 days. Toxin production was demonstrated in 25 of the cooked vegetables inoculated with proteolytic strains and in 13 products inoculated with non-proteolytic strains. The products which did not appear to support toxin production by proteolytic strains included garlic (pH 5.65), Brussels sprouts (pH 5.15) and tomatoes (pH 4.23) (Table 4.16). Carlin and Peck (1995) demonstrated a general relationship between growth and toxin

Table 4.16   Growth and toxin production by *C. botulinum* in cooked vegetables after storage at 30°C for 60 days, adapted from Carlin and Peck (1995)

Product	Toxin production		pH
	Proteolytic types A and B*	Non-proteolytic types B, E and F*	
Mushroom	2/2	2/2	6.39
Spinach	2/2	2/2	5.82
Potato	2/2	2/2	5.73
Bean sprouts	2/2	2/2	5.72
Garlic	0/5†	0/2†	5.65
Asparagus	2/2	2/2	5.39
Tomato + KOH (potassium hydroxide)	5/5†	0/2†	5.36
Courgette	2/2	1/5†	5.28
Green beans	5/5†	0/2†	5.21
Fennel	2/2	0/2†	5.18
Brussels sprouts	0/2†	0/2†	5.15
Carrot	4/5†	0/2†	4.99
Onion	2/2	0/2†	4.98
Green pepper	4/5	0/2†	4.95
Mange-tout	1/2†	0/2†	4.78
Tomato	0/2†	0/2†	4.23

Inoculum: 10^3 spores/g.
11 strains of non-proteolytic, five strains of proteolytic *C. botulinum*.
* Number of packs toxic/number tested.
† No evidence of visible signs of growth within 60 days.

production with the pH of the product for non-proteolytic and proteolytic strains. In general, the time to visible growth (where this occurred) by proteolytic strains in the puréed vegetables ranged from one day in mushrooms to 14–34 days in onion and the lower the pH the longer the time to visible growth. In common with many other studies, there were some products that showed no signs of visible growth but became toxic; these included green beans, celery and carrot. That pH was the predominant controlling factor in tomatoes was demonstrated by elevating the pH of tomatoes to 5.36 with potassium hydroxide after which toxin was produced in 5/5 samples after 60 days by proteolytic strains.

Products in which toxin was not produced by non-proteolytic strains were generally those close to or below pH 5.0, such as tomatoes, mange-tout, onion, carrot, turnip and parsnip. However, toxin was also not detected in garlic (pH 5.65) or tomatoes with elevated pH (pH 5.36).

It is evident that vegetable in oil products are not formulated to achieve safety are at risk with respect to potential for growth and toxin production by *C. botulinum*. Sufficient data exists from the various challenge test studies to demonstrate the potential for such products to cause botulism outbreaks. Although it may be possible to demonstrate the safety of particular vegetable in oil products by conducting challenge tests, it is clear from the conflicting reports in the literature that the 'natural' inhibitory effect of certain vegetables is variable and not especially reliable. Therefore, it is recommended that all products to be stored in oil at ambient temperatures made with herbs, spices and/or vegetables are processed to include a recognised *C. botulinum* controlling factor that ensures their safety. Such recognised factors include a pH value of 4.6 or less or a salt content of 10% or greater in the aqueous phase of the non oil components or a water activity of 0.94 or below in these components, achieved by the use of drying or humectants.

The high number of botulism outbreaks throughout the world caused by inadequately heat processed canned vegetables are testimony to the potential hazard associated with botulism from vegetables. Vegetables stored in oil at ambient temperatures do not receive the high heat processes given to canned products and many are being sold without clear evidence to support their safety. It is evident that many processors of such products are unaware of the risk they are presenting to the consumers of their products and it is incumbent on the food industry and those responsible for the establishment of legislation and enforcement of such legislation to ensure such processors are either stopped or made aware of the need to comply with some very basic principles of food safety; as a minimum this should include the application of hazard analysis based approaches to determining the appropriate safety controls for their products.

In addition, the multitude of recipes recommended by cookery books, leaflets and even television programmes advocating the home production of vegetables in oil make no mention of the potential hazard associated with this most dangerous of pathogens. Clearly, much education of these groups is also necessary (LaGrange Loving, 1998).

REFRIGERATED FRESH SALADS AND VEGETABLES

Traditionally, fresh vegetables were predominantly sold loose with little processing except for harvesting, transportation and retail handling. Shelf lives were governed by visual deterioration of the product and were consequently very short. With the advent of new processing techniques and a greater understanding of the factors contributing to the organoleptic quality of fresh vegetables together with the availability of modern packaging and gas packing systems, fresh vegetables can now achieve significantly longer shelf lives. Instead of being just harvested and sold directly, many salad vegetables are washed and prepared to be sold as ready-to-eat or ready-to-cook salads and, instead of being sold loose, many are now packed and presented in a more appealing manner. Although the advent of this new technology and marketing has distinct benefits in relation to the keeping quality of fresh produce, it has also brought with it some additional hazards afforded by extending the time over which pathogenic microorganisms may grow and by inhibiting the normal microflora of vegetables from spoiling the product by natural means. Prepared vegetables, particularly salad produce, have been a major growth market in the last decade and significant growth continues.

Description of process

Prepared vegetables can consist of basic produce, such as carrots, potatoes, broccoli, cabbage or other field vegetables, which are harvested, washed to remove soil and then packed in plastic packaging prior to transportation, often under refrigeration, to retail stores for sale (Figure 4.3). Such products may be sold under refrigeration but they may also be sold under ambient conditions. Shelf lives vary with each commodity but can range from several days for carrots to over a week for potatoes. The products in these categories are usually washed prior to consumption with most being consumed after cooking although some may be consumed raw, e.g. carrots, broccoli, etc. Products receiving similar, simple processing include many of the salad items, such as lettuce, cucumber, mushrooms, spring onions, radishes, which are harvested and may or may not be washed and some of which may be packaged in plastic packaging, overwrapped and sold with very short shelf lives of a few days. Again, they are usually displayed under chilled conditions but they are also frequently sold from ambient retail cabinets. Such products are usually washed by the customer prior to consumption although some may be cooked prior to eating.

Many of the products in both of these categories are not really considered

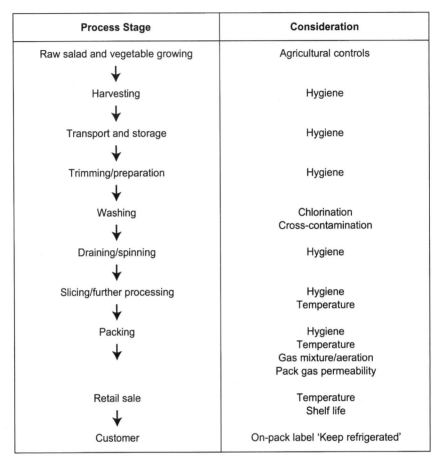

Process Stage	Consideration
Raw salad and vegetable growing	Agricultural controls
Harvesting	Hygiene
Transport and storage	Hygiene
Trimming/preparation	Hygiene
Washing	Chlorination Cross-contamination
Draining/spinning	Hygiene
Slicing/further processing	Hygiene Temperature
Packing	Hygiene Temperature Gas mixture/aeration Pack gas permeability
Retail sale	Temperature Shelf life
Customer	On-pack label 'Keep refrigerated'

Figure 4.3 Process flow diagram and technical considerations for typical prepared salads and vegetables.

to be prepared produce but are reviewed in this section because of the tendency for many of them now to be pre-packaged, in comparison to loose display that was more common in previous years.

The products most commonly recognised in this area are the prepared ready-to-eat salads and vegetables. The processes by which these are manufactured vary but they all start with field harvesting of the salad crop and storage (usually chilled) prior to delivery to the processing unit. The raw salad and vegetable ingredients are then prepared by, for example, trimming the outer leaves and removing the core of materials such as cabbage or lettuce. Other materials, such as carrots or onions, may be peeled using automated peeling machines. Most prepared produce is washed after the initial preparation stages in chlorinated water. The water is removed by draining or spinning and then the product is cut and assembled into the packs. The pack may be gas flushed or may just contain a normal gaseous atmosphere; it is then heat-sealed and chilled.

Prepared salads and vegetables are given short shelf lives of 4–6 days and this is primarily due to the significant visual and organoleptic deterioration that occurs with these products rather than the potential for growth of contaminating pathogens. In some countries the products may be sold at ambient temperatures either in pre-packs or on market stalls as loose commodities where the shelf life is further restricted by the visual appearance of the product which, under these conditions generally deteriorates in 1–2 days.

Irrespective of the gaseous mixture used in the pack, pre-packed salads and vegetables, including raw vegetables, can become anaerobic very quickly due to continued respiration of the plant tissues with the concomitant production of carbon dioxide and depletion of oxygen.

Raw material issues and control

The main points of microbial contamination of the salad and vegetable raw ingredients are during growing and harvesting. While the adoption of good agricultural practices during these stages may play a role in reducing the extent of contamination, such practices are likely to have a limited effect in reducing the incidence of contamination with spores of *C. botulinum*. However, preventing the application of animal and human wastes directly onto growing crops together with minimising the amount of residual soil retained by the harvested crop are all important activities in the overall attempt to reduce the risk associated with these products.

In terms of active mechanisms to reduce contamination, at best, salads and vegetables which are to be packaged may receive a chlorinated water wash to reduce soil and dust loading. Such processing will achieve little more than perhaps a 1–2 log reduction in microbial contamination when using wash water containing 100 ppm free chlorine (Beuchat, 1992) and cannot be considered to be capable of delivering a raw material free from bacterial contaminants. As the microflora of the plant will be derived primarily from the soil, although airborne and waterborne contaminants may also be present, a high loading of microorganisms, including bacterial spore-formers, can be expected on these products. *C. botulinum* will inevitably be a constituent component of the raw vegetable microflora from time to time and even with the application of good standards of agricultural practice it is not possible to exclude its presence. The presence of the organism in the soil has already been reviewed (Table 1.5) and it is clear that it can be present at high frequencies, with types A and B predominating.

With any raw material receiving a wash, it is important to minimise spread of contamination from a discrete location on one vegetable to an entire batch. This is very difficult to control in any washing system where continuous washing occurs and where the water may be recirculated. Although free spores of *C. botulinum* can be inactivated by the low levels of chlorine used in wash water (often 50–100 ppm free chlorine for prepared salads), inactivation of the organism attached to the vegetable is likely to be less effective.

When washing salads and vegetables, it is essential to prevent build-up of organic debris in the wash water itself because it may inactivate any residual chlorine present or allow proliferation of contaminants, making the washing process more of a hindrance than a help. The primary means of preventing contamination is by regular changing of process water and particularly when build-up of soil is apparent. Monitoring chlorine levels is also important and there should always be free available chlorine present, as this is the component active against the microbial contaminants.

The incidence of *C. botulinum* in vegetables has been reviewed recently by Dodds (1993b), with reported incidence varying from 0% to 43%. In a survey of vegetables from the USA conducted by Meyer and Dubovsky (1922b), 29 out of 189 samples were positive for types A and B. Products from which the organism was isolated included beans, carrots, potatoes, corn, celery, olives and asparagus. Dodds (1993b) also reported that other surveys found the same types in samples of cabbage, onions, tomatoes, dill and parsley. Clearly, therefore, *C. botulinum* will be present in these materials on occasion.

Process issues and control

Products of this nature receive very little processing and little can be done to make the product itself any more stable in relation to the growth of *C. botulinum*. Products which are cut/sliced before packing are potentially a greater risk than those harvested and sold whole. This is due to the localised release of nutrients and moisture made available by slicing to any contaminating microorganisms for growth. Although water should be reduced to a minimum after washing by spinning or draining, a washed product is inevitably packed in a damp condition.

The principal hazard associated with sliced or whole vegetables arises as a result of packing. Anaerobic environments develop due to the continued respiration of the vegetable within the closed pack which depletes oxygen levels. Alternatively, such anaerobic environments may be created artificially by introducing a modified atmosphere during packing of the product and prior to sealing. Even though the packaging can be designed to allow gaseous exchange with different transfer rates of oxygen and carbon dioxide, it is very difficult to prevent some parts of the produce from becoming exposed to anaerobic conditions. Such conditions clearly increase the potential for growth and toxin formation by *C. botulinum*. Although growth and toxin formation could equally occur in fresh vegetables not packaged anaerobically, i.e. in anaerobic 'pockets' within the product, the extended life that these packaged products are given because of the reduced rate of physical and microbiological deterioration provides greater opportunities for the development of hazards such as strains of *C. botulinum* without the usual signs of spoilage.

Virtually any fresh vegetable has the potential to allow growth and toxin production by *C. botulinum*. Some vegetables may have naturally low pH values (pH < 5.0) which could prevent the growth of some strains, particularly the non-proteolytic types, and this may be an important consideration in establishing the shelf life of such products. Products intended to be displayed under refrigerated conditions (< 8°C) will exclude the potential for growth of proteolytic strains of *C. botulinum* and such products are primarily at risk from non-proteolytic strains due to their ability to grow at refrigeration temperatures. It is, however, essential that if intended to be stored under chilled conditions for reasons of safety, these products should be labelled clearly 'Keep refrigerated' in large, bold and legible print. This is also often accompanied by a durability indicator such as a 'Use by' date and information 'Consume within use by date'. Prepacked vegetables and salads destined to be stored at ambient temperatures are susceptible to the growth of both proteolytic and non-proteolytic strains of *C. botulinum*.

As packing under conditions that allow anaerobic environments to develop quickly may play a role in increasing the potential for *C. botulinum* growth, it is possible that venting the pack to allow diffusion of oxygen and carbon dioxide may reduce the extent of anaerobiosis. Punching holes in vegetable packs to reduce the amount of anaerobiosis has been shown to prevent the growth of *C. botulinum* in some products (Table 4.17). However, this may in itself compromise the quality of the product, rather defeating the point of packaging in the first place. Importantly, while venting the pack will play a role in reducing the rate at

Table 4.17 Growth and toxin production by *C. botulinum* in overwrapped mushrooms, with and without holes in the packaging, adapted from Sugiyama and Rutledge (1978)

Holes in pack	Number of packs tested	Number toxic	Oxygen content (%)
0	28	28	0.9–2.5
1	123	0	1.9–6.6
2	47	0	4.9–7.3

Stored at 24–26°C for six days.
Inoculum: 5 × 10^5 heat shocked (80°C, 15 minutes) spores per mushroom. Four type A and four type B strains.

which the vegetables may become toxic, it cannot be guaranteed that it will not occur given sufficient time.

Petran *et al.* (1995) studied the potential for toxin formation in cut lettuce and sliced cabbage. Fresh cut romaine lettuce and shredded cabbage were inoculated with 10^4 heat shocked spores of *C. botulinum* (5 strains of proteolytic type A, 4 strains of proteolytic type B and 3 strains of non-proteolytic type B). Products were stored in polyester bags which were either non-vented or vented with four 0.25 inch holes. Products were stored at 4.4°C, 12.7°C and 21°C for 28 days. Toxin was not detected in any product stored at 4.4°C or 12.7°C in the 28 days. At 21°C, toxin was detected in all three replicate samples of non-vented shredded cabbage after 10 days' incubation, although the product was described as inedible due to a brown, slimy appearance. No samples of cabbage in vented packages allowed toxin formation, which may have been due to the presence of oxygen or more accelerated spoilage and lowering of pH, although the latter was not reported in the paper. All replicates of romaine lettuce in non-vented packs were positive for toxin after 17 days at 21°C, although again, the products were considered to be inedible. However, samples of lettuce in vented packs were also found to contain toxin after 28 days at 21°C. This was attributed to the development of localised anaerobic conditions in the pack possibly due to moisture accumulation as a result of product deterioration. The product was naturally considered to be severely spoiled.

An alternative way of making the products safe would be to employ techniques designed to inhibit the growth of the organism during the shelf life. The addition of salad dressings, thoroughly coating the vegetable materials, can significantly reduce the pH of the product making them inhibitory to growth. Dressings, which are usually mayonnaise based bringing the pH to < 5, preclude the growth of non-proteolytic *C. botu-linum*, but of course, the addition of dressings substantially changes the nature of the product.

Alternative approaches might be to add organic acids by dipping the salad components. However, pre-packed sliced vegetables are highly suscep-tible to flavour changes introduced by the addition of organic acids and addition of inhibitory compounds to whole fresh packed vegetables is not possible if natural organoleptic quality is to be maintained.

Final product issues and control

Fresh vegetables packed in plastic packaging are susceptible to con-tamination, growth and toxin production by strains of *C. botulinum*. The safety of these products can only be maintained by a full consideration of the potential for the growth of *C. botulinum* given the anticipated storage conditions of the product, i.e. times and temperature.

The incidence of *C. botulinum* in fresh salads and vegetable products has been studied by several authors (Table 4.18). Lilly *et al.* (1996) studied large numbers of commercially prepared salads, onions, broccoli, stir-fry vegetables, etc. and found only four samples to be contaminated. Three samples contained type A *C. botulinum* and one sample contained both type A and type B. The positive products included shredded cabbage (type A), chopped green pepper (type A), salad mix (type A) and escarole salad mix (types A and B).

In a survey by Gibbs *et al.* (1994) for psychrotrophic strains of *C. botu-linum*, samples of carrots (48), lettuce/mixed salad (71) and watercress (20) were enriched with small volumes of cooked meat broth containing glucose and incubated at 10°C for three weeks. Based on the absence of toxin production under these conditions, none of the samples were found to contain psychrotrophic strains of the organism.

Based on the assumption that conditions can become anaerobic in any gas pack, the only real botulinum-controlling factor is shelf life (with the exception of those products which are naturally acidic). Therefore, any chilled, fresh-prepared salad or vegetable should either have a pH of 5.0 or less or be assigned a shelf life of 10 days or less if stored under refrigerated conditions, e.g. < 8°C. This is consistent with the advice given in the UK by the Advisory Committee on the Microbiological Safety of Food (1992) for vacuum-packed, chilled products and, although primarily intended

Table 4.18 Survey of the incidence of *C. botulinum* in commercially prepared fresh vegetables and salad products

Product	Sample number	Test weight	Number positive	Reference
Shredded cabbage	337	454 g	1 (type A)	Lilly *et al.* (1996)
Chopped green pepper	201		1 (type A)	
Shredded coleslaw	72		0	
Carrot	7		0	
Onion	4		0	
Broccoli	3		0	
Mixed vegetables	90		0	
Salad mix (Italian)	24		1 (type A)	
Salad mix (oriental)	35		0	
Stir-fry vegetables	54		0	
Caesar salad	53		0	
Salad mix (iceberg)	58		0	
Salad mix (escarole)	58		1 (types A and B)	
Salad mix (butter lettuce)	64		0	
Salad mix (green leaf lettuce)	58		0	
Carrots	48	100 g	0	Gibbs *et al.* (1994)*
Lettuce/mixed salad	71	100 g	0	
Watercress	20	75 g	0	

*Survey for psychrotrophic strains only.

for processed fish and meat products, this advice can equally be applied to salads and vegetables. Likewise, those products sold under ambient conditions should have shelf lives restricted if the potential for growth of *C. botulinum* is demonstrated.

While some of the products in this category are raw and should be fully cooked prior to consumption, e.g. potatoes, it is unwise to rely on cooking by the consumer to destroy any toxin that may have formed in a product of unsafe formulation or shelf life. It is critical to ensure that the products cannot become toxic in the normal shelf life of the product and this may necessitate clear use instructions to the consumer. This is particularly relevant to chilled perishable products where the consumer may not be aware of the critical role temperature may be playing in ensuring safety. Where it is important, both the shelf life and storage conditions must be clearly marked on the product to be readily seen by the consumer.

That *C. botulinum* can grow and produce toxin on fresh salads and vegetables has been shown by many researchers. Austin *et al.* (1998) demonstrated growth and toxin production in a variety of fresh-cut and prepared salad products stored under modified atmospheres where the final pack oxygen levels were < 1.5% and the carbon dioxide levels were above 15% (usually 25–45%) (Table 4.19). When stored at temperatures of 15°C or higher, proteolytic strains of types A and B were capable of toxin production in salad stir-fry after 21 days (15°C storage, 10 spores/g inoculum) and 11 days (25°C storage, 100 spores/g inoculum). In romaine lettuce the toxin was detected after 9 days at 25°C from an inoculum of 100 spores/g and in mixed salad under the same conditions toxin was detected after only 7 days. Although toxin production was generally accompanied by extensive decay, in some cases the only physical/orga-noleptic change was described as soft. In the same study, non-proteolytic strains did not grow in any product at 5°C, with the exception of butternut squash where toxin was produced after 21 days. However, at 15°C and 25°C non-proteolytic strains of type B and type E inoculated at 1000 spores/g produced toxin in mixed, prepared salad after 14 days and 4 days respectively and were accompanied by only moderate browning of the product. Although such high levels of contamination would not normally be anticipated, this demonstrates the potential hazard associated with unrefrigerated display of some types of prepared salad.

Larson *et al.* (1997) inoculated packs of lettuce, cabbage, broccoli, carrots and green beans with a mixture of heat shocked spores of seven pro-teolytic and three non-proteolytic strains of *C. botulinum* types A, B and E. Products were inoculated at levels of c. 100 spores/g and the products were stored in packaging of two different oxygen and carbon dioxide transmission rates at 4°C, 12°C and 21°C. Samples were tested for *C. botulinum* toxin prior to onset of spoilage and after gross spoilage had occurred. Botulinum toxin was detected in two out of six grossly spoiled lettuce samples after six days' storage at 21°C, one from each type of packaging (Table 4.20). No toxin was detected in lettuce stored for up to 50 days at 4°C or in samples incubated at 12°C for 7 or 13 days, even though in the latter sample the product was grossly spoiled. The toxic packs had final pH values of 4.48 and 4.88 while the oxygen content was < 0.03% and the carbon dioxide c. 30–40%. The only other product in which toxin was produced was broccoli, where samples which were overtly spoiled after storage at 12°C for 9 days were found to be toxic. The final pH of these packs was 6.58, the oxygen content 0.4–0.8% and carbon dioxide c. 10–11%. Broccoli stored at 21°C for 3 and 7 days, also deemed to be 'spoiled' and 'grossly spoiled' respectively, was also toxic. pH values in final product varied from 5.72 to 6.62, oxygen content 0.75–1.68% and

Table 4.19 Toxin production by *C. botulinum* in fresh-cut vegetables stored under modified atmosphere, adapted from Austin *et al.* (1998)

Product	Inoculum (spores/g)	Type	Temperature (°C)	Time to toxin production (days)
Stir-fry salad	1000	Non-proteolytic	5	Not toxic*
	1000	Non-proteolytic	10	Not toxic
	100	Proteolytic	15	Not toxic
	10	Proteolytic	15	21
	100	Proteolytic	25	11
Romaine lettuce	1000	Non-proteolytic	5	Not toxic
	1000	Non-proteolytic	10	Not toxic
	100	Proteolytic	15	Not toxic
	10	Proteolytic	15	Not toxic
	100	Proteolytic	25	9
Onion	1000	Proteolytic	25	6
Mixed salad	1000	Non-proteolytic	5	Not toxic
	1000	Non-proteolytic	10	Not toxic
	1000	Non-proteolytic	15	14
	1000	Non-proteolytic	25	4
	100	Proteolytic	15	Not toxic
	10	Proteolytic	15	Not toxic
	100	Proteolytic	25	7
Butternut squash	1000	Non-proteolytic	5	21
	1000	Non-proteolytic	10	7
	100	Proteolytic	15	14
	10	Proteolytic	15	14
	100	Proteolytic	25	3

Final pH of toxic samples ranged from 4.2 to 6.7.
*Sampled up to a maximum of 21 days.

carbon dioxide 11–13%. Although oxygen content, carbon dioxide content and pH were found in many other products examined to be similar to those in the toxic vegetables, toxin was not detected in the other products. In addition, lettuce samples pierced to simulate abuse did not yield any toxic packs. This work indicated that toxin was not detected in any of the vegetables tested prior to the onset of visual spoilage.

In contrast to many of these findings Hao *et al.* (1998) recently found no toxin production by a 10-strain cocktail of proteolytic strains of *C. botulinum* inoculated at c. 10^2 spores/g into shredded cabbage and lettuce packed in high oxygen and low oxygen transfer films. Product was stored

Table 4.20 Storage conditions for lettuce and broccoli allowing growth and toxin formation by *C. botulinum*, adapted from Larson *et al.* (1997)

Product	Incubation temperature (°C)	Incubation time (days)*	Number of toxic packs	Conditions in pack at end of incubation			
				pH	Oxygen (%)	Carbon dioxide (%)	Organoleptic condition
Lettuce	21	6	1/3	4.48	<0.03	32.7	Gross spoilage
		6	1/3	4.88	<0.03	40.7	Gross spoilage
Broccoli	12	9	1/3	6.58	0.4	10.8	Gross spoilage
		9	2/3	6.58	0.8	9.9	Gross spoilage
	21	3	3/3	6.62	1.19	13.3	Spoilage
		3	3/3	6.58	0.75	11.4	Spoilage
	21	7	3/3	5.8	1.68	13.2	Gross spoilage
		7	3/3	5.72	1.76	11	Gross spoilage

* Results are from packs inoculated with spores and packaged in film of two different gas permeabilities.
All products stored at 4°C, 12°C and 21°C and tested for toxin prior to spoilage, after spoilage and after gross spoilage. All other combinations were negative for toxin production prior to product spoilage and after signs of gross spoilage.

at 4°C, 13°C and 21°C for up to three weeks (cabbage) or four weeks (lettuce). The pH decreased markedly in all products reaching pH < 4.6 (initial pH 6.4) in most samples, although this decrease occurred most rapidly in products stored at higher temperatures. Oxygen levels became depleted within several days (c. 20% to < 2%) in all types of packaging although this occurred fastest in lettuce, attributed to a higher respiration rate. Carbon dioxide levels increased more rapidly in low oxygen transfer film and the concentration also increased to higher final levels (from c. 2% initially to 40-60%). Clearly, proteolytic *C. botulinum* would not be expected to grow at 4°C and would grow fairly slowly at 13°C but it is not clear why it did not grow to produce toxin at 21°C. However, it is probable that the development of other microbial species together with the rapid decrease in pH may have created local conditions capable of inhibiting growth of *C. botulinum*. Indeed, this is the very principle on which many traditional fermented vegetables are prepared using 'natural' microflora.

In a further publication, Hao *et al.* (1999) reported on studies of the growth of a 10-strain mixture of proteolytic strains of *C. botulinum* (10^2 spores/g) in broccoli florets, shredded carrots and green beans. Products were stored at 4°C, 13°C and 21°C for up to 28 days in packaging of different oxygen transmission rates; $3000 \, cm^3/m^2/24 \, h$ for carrots, $6000 \, cm^3/m^2/24 \, h$ for beans, $7000 \, cm^3/m^2/24 \, h$ for carrots and broccoli and $16000 \, cm^3/m^2/24 \, h$ for broccoli and beans. The packaging type affected product deterioration although this was highly variable for different product types and storage temperatures. Botulinum toxin was only produced in samples of broccoli stored at 13°C after 21 days (oxygen transmission rate of $7000 \, cm^3/m^2/24 \, h$) and at 21°C after 10 days (oxygen transmission rates of $7000 \, cm^3/m^2/24 \, h$ and $16000 \, cm^3/m^2/24 \, h$). Toxin was not produced in any other product under any other condition. All samples of broccoli were visibly spoiled at the point of toxin production.

C. botulinum can equally grow and produce toxin in less conventional packaged vegetable products such as potatoes. Solomon *et al.* (1994) inoculated sliced peeled potatoes with spores of *C. botulinum* types A and B and incubated them under vacuum at room temperature (22°C). *C. botulinum* grew and produced toxin in the raw potatoes in three days which coincided with the visual deterioration of the product (Table 4.21). Products were considered acceptable on day 2 when no toxin was detected. Products dipped in sulphite were found to be organoleptically acceptable up to day 6 whereas toxin was detectable (type A) after only four days. This finding was reinforced in a similar study (Solomon *et al.*, 1998) with sulphite and non-sulphite treated potatoes stored under

Table 4.21 Growth and toxin production by *C. botulinum* in raw, untreated sliced potatoes, adapted from Solomon *et al.* (1994)

Type A

Inoculum (spores/g)	2 days		3 days	
	Toxic (no. toxic/no. tested)	Quality	Toxic (no. toxic/no. tested)	Quality
550/g	0/2	Acceptable	3/5	Marginally acceptable
5000/g	0/10	Acceptable	19/20	Marginally acceptable/unacceptable
—	—	—	—	—

Type B (proteolytic)

Inoculum (spores/g)	2 days		3 days	
	Toxic (no. toxic/no. tested)	Quality	Toxic (no. toxic/no. tested)	Quality
1220/g	0/5	Acceptable	0/5	Acceptable/marginally acceptable
5500/g	0/10	Acceptable	2/10	Unacceptable
12000/g	0/5	Acceptable	0/5	Unacceptable

Five-strain mixture of type A spores or proteolytic type B spores. Vacuum packed and stored at 22°C.

Table 4.22 Growth and toxin production by *C. botulinum* in raw, sulphite-treated sliced potatoes, adapted from Solomon *et al.* (1994)

	Type A					Type B				
		Toxin production after storage period (number of samples toxic/number of samples tested)					Toxin production after storage period (number of samples toxic/number of samples tested)			
Inoculum (spores/g)	SO$_2$ (ppm)	4 days*	5 days*	6 days*		Inoculum (spores/g)	SO$_2$ (ppm)	4 days*	5 days*	6 days*
3000–3750/g	136–174	4/7	7/10	8/8		3000–3050/g	117–179	0/7	4/11	—
5500–6000/g	144–200	4/8	4/4	3/4		4650–6500/g	138–147	0/6	3/11	0/4

Five-strain mixture of type A spores or type B spores. Vacuum packed and stored at 22°C.
*Quality judged to be acceptable at all time points.

modified atmosphere packing (30% nitrogen, 70% carbon dioxide). Untreated potatoes stored under these conditions at 22°C became toxic after 4–5 days which coincided with the description 'unfit for consumption'. In contrast, those treated with sulphite became toxic after four days and were considered organoleptically acceptable up to seven days (Table 4.22).

Clearly, the implications of adding preservatives that maintain product quality but have no adverse effect on contaminating pathogens is a potentially dangerous development and the management of such hazards brought about by technical approaches to extend the shelf lives of these products needs to be built into the safety controls for the product.

Although many of the reported studies of growth of *C. botulinum* in these product types have involved inoculum levels higher than might be expected in practice, it is important to accept that the organism will be present on the product components and, when present, however low the number, can grow under many of the packaging and storage conditions applied. If the organism grows, then toxin production is possible. Thus, *C. botulinum* must be considered in the hazard analysis conducted during the development and production of these types of product.

REFRIGERATED, COOKED FISH PRODUCTS (INCLUDING SMOKED FISH)

Fish products do not have a particularly good safety record in relation to botulism. Many outbreaks have been recorded associated with these products, although most documented cases relate to fish products processed using traditional methods often involving natural fermentation. Commercially cooked and smoked fish products have been associated with relatively few outbreaks of botulism but outbreaks involving other fish products are testimony to the high frequency with which *C. botulinum* is likely to be present in the raw material. Products discussed in this section range from simple cooked fish and shellfish such as prawns and surimi to the more complex processed products such as cold smoked salmon and hot smoked mackerel. Cooked fish and shellfish are consumed by all sectors of society whereas smoked fish, particularly smoked salmon and trout, tend to be consumed more by adult populations and smoked fish roe (eggs – caviar) tends to be consumed by more affluent sectors due to its high price, although less expensive varieties from fish more common than sturgeon are available.

Description of process

Cooked fish and shellfish

Cooked fish products are made from any type of fish or shellfish by a variety of processing techniques (Figure 4.4).

Shellfish such as prawns may be farmed or harvested at sea and stored on ice. The shellfish is usually cooked in a factory although cooking on board ship is also practised, particularly on ships at sea for extended periods. Shellfish are transported to processing sites chilled, sometimes on ice and then washed prior to cooking. They are often cooked by immersion in hot water or they may be cooked by passing under a steam/hot-water spray on a conveyor, after which they are chilled rapidly by cold water immersion or blast chilling. Shellfish may be packed in brine. Shellfish are not normally cooked in pack and they often receive further processing following the cook, particularly if the shells have to be removed. With the exception of some molluscan shellfish, e.g. cockles and mussels, shellfish are rarely cooked to temperatures in excess of 75–80°C as the meat becomes very tough ('rubbery') if cooked for too long or at too high a temperature.

Bivalve molluscs such as cockles and mussels are often cooked at temperatures in excess of 90°C for 90 seconds if there is the potential for

Process Stage	Consideration
Fish farming	Health Cleanliness
or	
Sea fishing	Hygiene
↓	
Storage, if applicable	Hygiene Temperature/time
↓	
Slaughter, gutting, filleting	Hygiene
↓	
Transport, if applicable	Hygiene Temperature/time
↓	
Storage	Hygiene Temperature/time
↓	
Packing (if applicable)	Seal integrity
↓	
Cooking	Temperature Time
↓	
Cooling	Temperature Time
↓	
Transfer to high-risk area (if applicable)	Hygiene
↓	
Packing (if applicable)	Hygiene Temperature Time
↓	
Storage	Temperature Time
↓	
Distribution	Temperature Time
↓	
Retail sale	Temperature Shelf life
↓	
Consumer	On-pack label 'Keep refrigerated'

Figure 4.4 Process flow diagram and technical considerations for typical cooked fish.

contamination with human enteric viral pathogens such as small round structured viruses (SRSV) and this heat process has historically been shown to prevent recurrence of outbreaks of viral gastroenteritis associated with these products. Typical heat processes for other shellfish achieve 70–72°C for less than two minutes.

After cooking and chilling, shellfish are packed into bulk packs which may be gas flushed for distribution to retail stores where they may be displayed on the fish counter, usually in open containers on beds of ice. Alternatively, they may be packed at the processor's into pre-packs or onto trays and overwrapped with film to be sold as pre-packaged units of fresh shellfish from retail outlets. Other pre-packs may be frozen and either sold frozen or defrosted and sold chilled.

It is customary for many fishmongers and fish stalls at seafront locations to sell shellfish loose for direct consumption and, under these circumstances, the cooked shellfish is scooped directly into small individual containers for immediate consumption.

Cooked, chilled shellfish is sold under refrigerated conditions with a relatively short shelf life dictated by the growth of spoilage bacteria rather than bacterial pathogens. Unless packed under modified atmosphere or cooked in pack, shellfish will have a shelf life of less than 10 days under refrigerated conditions. However, shellfish cooked in pack and those stored under modified atmospheres may have allocated shelf lives of more than 10 days.

Cooked fish receives similar processing to shellfish, except it often receives significant handling after catching and prior to cooking. Fish caught at sea is immediately stored on ice and may be gutted prior to shipboard storage or gutting may occur after landing. The fish is usually gutted and filleted prior to cooking and it may also have the skin removed. Fish may be cooked whole, coated in batter and crumbs and flash fried or it may be minced and reformed with other ingredients into different shapes and sizes prior to cooking. The cooked fish is usually blast chilled after cooking and is then packed. Few cooked fish products are cooked in pack, although some products such as surimi (minced, reformed and shaped fish and shellfish) may be cooked and cooled in hermetically sealed packs prior to distribution and sale. Cooked fish are allocated shelf lives of less than one week unless cooked in pack or packaged in modified atmosphere or vacuum packs when products may be given a shelf life in excess of two weeks under refrigeration conditions.

Some fish are included in recipe dish (ready meals) formulations and/or may be cooked by the sous-vide process. Any such variation must be taken into account during process hazard analysis.

Smoked ready-to-eat fish

Fish and shellfish are frequently smoked to add flavour to the products. This may be in the form of artificial liquid smoke added as an ingredient or it may be achieved by placing the raw material in smoking chambers where the smoke is produced by burning wood or wood shavings. Artificial smoke flavour is usually used for adding flavour to raw fish destined to be cooked although some fish may receive natural smoking at low temperatures. Hot smoked fish and shellfish are usually smoked after a brining or salting stage which is normally carried out under refrigerated conditions (Figure 4.5). Hot smoked fish products such as mussels or mackerel are cooked and smoked in a combined process, i.e. the temperature in the smoking room is elevated to the cooking temperature during the smoking process (hot smoking). Hot smoking usually involves placing the fish in kilns for several hours during which time the internal temperature of the fish reaches 65–75°C for approximately 30 minutes. After cooling, the fish is then packed, usually under vacuum, and then sold under refrigerated conditions with shelf lives of several weeks.

Some fish undergo a cold smoking process which contributes the smoke flavour but does little to destroy vegetative pathogens (Figure 4.5). Fish for cold smoking such as salmon and trout are usually farmed but may also be caught wild. The fish are killed at the farm and then transported on ice to processing plants where they are gutted, filleted and prepared as two sides for salting. Fish may be dry salted by adding the salt directly onto the surface of the fillets with the skin still attached, although some processors may immerse the fillets in a brine solution or even inject brine. To ensure equilibration of salt, the fish are stored under refrigerated conditions for several days and then they are transferred into the smoke rooms. The salting stage is designed, in most cases, to achieve a concentration of salt of 3.5% or more in the aqueous phase of the fish. Smoking is then carried out at ambient temperatures (25–30°C) using burning wood dust or wood chips for several hours. After smoking, the fish are chilled prior to packing as whole sides or after being cut into thin slices manually using hand held knives or slicers, although some automated slicers are now used for further processed products. The sides or slices are usually vacuum packed and retailed with a shelf life of 3–4 weeks or more at chill temperatures (<5°C). In some countries it is common practice to retail these products

Process Stage	Consideration
Fish farming	Health Cleanliness Fish meal quality
↓	
Slaughter, gutting, filleting	Hygiene
↓	
Transport	Hygiene Temperature Time
↓	
Storage	Hygiene Temperature Time
↓	
Trimming/halving	Hygiene Temperature Time
↓	
Salting	Distribution of salt Salt levels Hygiene
↓	
Storage	Hygiene Temperature Time
↓	
Smoking	Hygiene Temperature Time Type of smoke
↓	
Storage	Hygiene Temperature Time
↓	
Slicing (if applicable)	Hygiene
↓	
Packing	Hygiene
↓	
Distribution	Temperature
↓	
Retail sale	Hygiene Temperature Shelf life
↓	
Consumer	On-pack label 'Keep refrigerated'

Figure 4.5 Process flow diagram and technical considerations for typical ready-to-eat smoked fish.

by mail order and they are often sent via the postal system in insulated packs or with a cold pack for overnight delivery.

The products of greatest concern in this category with respect to the hazard of *C. botulinum* are those that are allocated extended shelf lives which includes fish/shellfish cooked in pack or stored under vacuum or modified atmospheres and minimally processed, smoked fish products.

Raw material issues and control

Fish and shellfish are known to be contaminated with *C. botulinum* although the frequency and types will vary according to the source of the material. Marine sediments are frequently found to contain the organism and the fish may become contaminated during its normal lifecycle or during catching or processing (Table 1.6). Fish may be contaminated on the external surfaces but it is most common to find the organism in the intestinal contents. The type most commonly associated with marine sediments and therefore also the fish is *C. botulinum* type E. As strains of this type can grow psychrotrophically, it is evident why chilled, long shelf life products of this nature are of particular concern with regard to the hazard of this organism. This is especially so for those products which receive processing insufficient to destroy the contaminant and where inherent factors preventing growth are inadequate and therefore not growth limiting in the food.

In a survey of aquatic environments including lakes, ponds, reservoirs, marshes, mud flats, streams, rivers and canals in Great Britain and Ireland, Smith *et al.* (1978) did not find any *C. botulinum* type A in 554 samples of mud. However, 35% of samples contained one or more other type with 30.1%, 3.4%, 1.1% and 2.7% being found to contain types B, C, D and E respectively and 2.3% containing more than one type. Each type found was demonstrated in both fresh and salt water environments. In a study of bottom deposits from the North Sea and Scandinavia, Cann *et al.* (1965) found little evidence of type E strains in the North Sea but, in contrast, samples from deposits around the Scandinavian coast were mostly found to contain the organism. In addition, of 130 fish intestine samples and 131 fish samples caught in UK coastal waters and inoculated into broth to determine toxin production, none were found to be positive indicating absence of toxin type E strains in the samples. Huss *et al.* (1974a) demonstrated a high incidence of *C. botulinum* type E in trout farms in Denmark. A total of 530 trout were examined with incidence varying from 5 to 100% in the winter and 85 to 100% in late summer. With very few exceptions, *C. botulinum* type E was the predominant isolate, although types A and B were also found on occasion. It was postulated that the high incidence of this organism in fish farms was partly contributed by the use of fish feed derived from minced marine 'trash' fish which may contain high numbers of *C. botulinum*. Some points associated with the control of the organism on the farm and during subsequent processing were noted by Huss *et al.* (1974b). High among these were careful gutting practices and associated hygiene precautions to prevent contamination of the gutted fish and subsequent fish handled on the same production line. In addition, the use of lime treatments in the fish farm was shown to be associated with a reduced incidence of the organism.

In a review of the significance of *C. botulinum* in fishery products, Eklund (1982) noted the widespread occurrence of *C. botulinum* (type E in particular) from freshwater and marine sediments in North America, southern California, the Atlantic Gulf coastal areas of the US, Canada, Russia and Scandinavia.

Hauschild (1989) extensively reviewed the literature then available regarding the incidence of *C. botulinum* in aquatic environments and reported levels ranging from < 1 to 2500 most probable number (MPN) per kilogram of sediment. Dodds (1993a) noted the often high incidence of the organism in fish and shellfish (Table 4.23), with incidences reaching 100% in one survey. The predominant type in most studies was type E.

The levels of *C. botulinum* can be very high, particularly in farmed fish sediments where the organism can enter via the fish food and proliferate in any dead fish. Indeed, botulism outbreaks in fish have occurred in salmon hatcheries.

Cann *et al.* (1975) reported the detection of types B, C, E and F in 10% of fish from trout farms (1400 samples). Huss *et al.* (1974a) found levels in fish from one trout farm ranging from 340 to 5300 spores per kilogram.

A recent survey of Finnish fish and fishery products for the type E botulinum neurotoxin gene using a polymerase chain reaction (PCR) method found 10%–40% positives in raw fish (intestine, surface and whole fish) and 4%–14% positive in fish roe (Hyytiä *et al.*, 1998). Isolates of *C. botulinum* were recovered from some PCR positive samples indicating the presence of this organism in European wild freshwater fish for the first time.

It is clear therefore from these surveys that *C. botulinum* will be present in fish and shellfish harvested from salt and freshwater environments and, while it may be possible to prevent extensive contamination by hygienic

Table 4.23 Incidence of *C. botulinum* in fish and shellfish selected from various surveys, adapted from Dodds (1993a)

Source	Sample	Incidence (percentage)	*C. botulinum* types
Lake Superior, USA	Fish	2	E
Gulf coast	Fish	4	B, C/D, E
	Shrimp	1	C, E
Alaskan coast	Salmon	5	E
South Baltic, Sweden	Fish	55	E
Skagerrak	Fish	0	—
UK, fish farm	Trout	10	B, C/D, E, F
Denmark, fish farm	Trout	64	A, B, E
Japan	Fish	5	E, F
Brazil	Oysters and shrimp	35	A, B, C/D, E, F

processing methods, it is not possible to eliminate its presence completely from the raw material. Therefore, like most raw materials, it must be assumed that *C. botulinum* will be present in raw fish and shellfish on occasion.

The single most important area where the incidence of *C. botulinum* in the raw material could be contained is the gutting stage, where good standards of hygienic processing must be maintained to prevent contamination of the fish with gut contents. Care must also be exercised to prevent opportunities for extensive cross-contamination of subsequent fish processed over common production lines in the gutting area and from common utensils such as knives.

It is also important to ensure that raw fish intended for cooking or for smoking is stored at temperatures preventing the growth of non-proteolytic *C. botulinum* prior to processing. This is usually achieved by chilled storage at very low temperatures (0-2°C) or freezing the fish as shelf life will be short (<6 days) under normal refrigeration conditions (5°C) due to growth of spoilage bacteria. It is becoming increasingly common practice to store raw material fish in bulk packs under modified atmosphere (MAP) to reduce the rate of microbial growth and extend the shelf life of the fish (this also being common practice for raw fish destined for retail sale).

It is important to recognise that MAP does not prevent the growth of *C. botulinum* in fish. Work, principally using raw fish, clearly demonstrates the potential for toxin formation under conditions of high carbon dioxide concentrations. Eklund (1982) demonstrated that toxin formation occurred in raw salmon inoculated with spores of *C. botulinum* type E after seven days' storage at 10°C. The level of spores required to produce toxin in seven days increased with increasing carbon dioxide concentrations: 10^3 spores/100 g for 60% CO_2 (25% O_2, 15% N_2) and 10^4 spores/100 g for 90% CO_2 (10% N_2). Importantly, in both cases, toxin was produced in the absence of microbial spoilage. In the same research, fish stored in air supported growth and toxin formation within seven days at 10°C from an initial inoculum of 100 spores/100 g. The product was, however, considered spoilt by the time toxin was formed. The application of strict temperature control (< 3°C), restriction of shelf life (10 days or less) or the addition of inhibitory compounds necessary to prevent the growth of *C. botulinum* in these raw materials are therefore essential.

Process issues and control

Cooked fish and shellfish

Fish and shellfish products intended to be cooked usually receive a heat process designed purely to achieve the destruction of vegetative pathogens. Shellfish are usually subject to cooking by immersion in hot-water tanks or by cooking on a conveyor passing under hot-water/steam sprays. Shellfish are very susceptible to losses in organoleptic quality if the heat process is excessive and, therefore, processing is designed to achieve product temperatures of 70°C–72°C for 30–60 seconds only. The cooked shellfish are rapidly cooled by cold-water sprays or immersion in cold water to < 5°C and, if necessary, shells are removed by automated machinery. The product is then packed into bulk packs for distribution to fishmongers or other retail outlets or it may be pre-packed in smaller units for sale directly to the customer.

Cooked fish are subject to similar heat processing although the heating methods can vary; they include steaming, frying or baking. Cooking processes are usually designed to achieve approximately 70°C for two minutes or an equivalent process. Products are usually cooled and packed into bulk packs or retail pre-packs.

The cooking process for both fish and shellfish is not designed to destroy spores of *C. botulinum*. The temperatures attained will have little effect on these contaminating pathogens. However, as most chilled, cooked fish and shellfish have shelf lives of less than seven days, the hazard presented by this organism is minimal providing effective temperature control is maintained. The principal hazard to these products comes from the non-proteolytic strains of *C. botulinum*, due to the potential for growth under extended chilled storage.

Cooked fish and shellfish products can support growth and toxin formation by non-proteolytic strains of *C. botulinum* but are usually not given sufficiently long shelf lives to allow such hazards to manifest themselves. This is because of the post-processing contamination with spoilage microorganisms that most products collect after cooking.

The cooked fish products of the greatest concern are those assigned long shelf lives and although vacuum-packed or MAP products can achieve longer lives, the presence of post-process contaminants even in these products will usually restrict shelf lives to < 15 days, if other inhibitory factors are absent.

To overcome this contamination issue, some processors package the fish or fish mixture in hermetically sealed packs and then apply the cook, e.g. sous-vide processes. Vegetative microorganisms are destroyed and the product is cooled without the potential for post-process contamination. As a result, such products can achieve shelf lives of several weeks at chill temperatures without spoiling. However, without the application of a cook sufficient to destroy spores of non-proteolytic *C. botulinum*, such products are a significant hazard. Therefore, these cook-in-pack fish products should be cooked to temperatures in excess of 90°C for 10 minutes to reduce the hazard to an acceptable level. This process is recommended by the UK Advisory Committee on the Microbiological Safety of Food (ACMSF) as the minimum for vacuum packed and MAP products with chilled shelf lives of > 10 days, so that a minimum of a 6 log reduction in the spores of non-proteolytic *C. botulinum* is achieved. However, studies have shown that this process may not always deliver such a reduction with all strains of non-proteolytic *C. botulinum* in all situations. For example, Peterson *et al.* (1997) studied the effect of heat on the destruction of *C. botulinum* type B spores in picked Dungeness crab meat. A mixture of three non-proteolytic strains was inoculated into the crab meat to give contamination levels of 10^6 spores per 30 g of packed crab meat. Packages were pasteurised at 88.9°C, 90.6°C, 92.2°C and 94.4°C. This resulted in *D* values of 12.9 minutes, 8.2 minutes, 5.3 minutes and 2.9 minutes respectively. To achieve a 6 log reduction in spores of the test strains would have required a heat process of nearly 60 minutes at 90°C, significantly longer than the 10 minutes recommended by the UK

Advisory Committee on the Microbiological Safety of Food. Nevertheless, the process recommended by the Advisory Committee would still be expected to result in a safe product when considering the likely contamination levels in the fish raw material.

If products are cooked in pack to destroy non-proteolytic *C. botulinum*, it is essential to assure the pack seal integrity to prevent re-contamination of the fish after processing. This most commonly occurs during cold water cooling as the hot fish contracts and draws a vacuum. Pack seal integrity checks are an essential element of product safety and measures employed include examining packs for evidence of vacuum loss. Care must be exercised however when using such simple measures as indicators of poor seal integrity, particularly immediately after cooking as any minor loss in seal or 'pin hole' defects may not result in loss of vacuum immediately but only after several hours. Ensuring seal integrity must be a major design requirement of the packing machinery and assured by effective operation and maintenance of the equipment. This must obviously go hand-in-hand with the use of appropriate strength packaging and sealing systems.

As with the cooking of any product, if the cooking stage is defined as a critical control point of the process, it is essential that the process is validated to ensure it is capable of achieving the correct temperature for the correct time in all parts of the product in all areas of the cooking equipment. All factors contributing to effective heat processing must be taken into account and should include, as a minimum, heat distribution in the oven, size of the fish portions to be cooked, temperature of the heating equipment, duration of the cook and volume of the product in the oven/heating equipment (maximum fill). All such factors play a major role in achieving the required process and must be validated under worst case conditions. Equipment used to monitor temperature and time must be accurate and regularly calibrated and, once established, the process should be monitored using continuously recording temperature probes and timing devices, preferably with chart recorders and alarms indicating loss in temperature or insufficient belt speed/process duration.

For cooked fish, especially that which is not given a non-proteolytic 'botulinum cook', it is essential to cool the product rapidly to prevent germination and growth of surviving spores and, as the size of most fish products is not very large, a temperature of 5°C should be achieved well within four hours.

Alternative strategies for achieving safety of cooked fish products could be to either acidify or salt the fish prior to or after processing. In such circumstances it is essential that the process for addition of the inhibitory agent allows equilibration to occur throughout the fish so that the pH or salt level is inhibitory throughout the product. That this is achieved should be checked as part of normal process control checks of the pH or salt content during the process.

In principle, products receiving a non-proteolytic 'botulinum cook' can be safely stored under refrigeration for long periods of time without compromising safety. However, it is important to note that as the spores of proteolytic strains of *C. botulinum* will remain unaffected by the pasteurisation processes employed to destroy non-proteolytic strains, any temperature abuse during the shelf life may result in an unsafe product. It is essential that products cooked in pack and given extended shelf lives under chilled conditions should be prominently labelled 'Keep refrigerated' in an attempt to prevent outbreaks occurring from surviving spores, other than non-proteolytic *C. botulinum*.

Smoked ready-to-eat fish

Fish products undergoing a natural smoking process are usually salted or brined prior to exposure to the smoke. During this stage of the process it is critical to ensure sufficient salt is present in the aqueous phase of the fish to prevent germination and subsequent growth of *C. botulinum*, both in the smoking stage and subsequent chilled shelf life storage of the product. Salting of fish is usually achieved by dry salt application or by soaking in brining solutions under chilled conditions for several days. Brine injection processes may also be used. In order to achieve complete inhibition of non-proteolytic strains of *C. botulinum*, it is essential to achieve salt concentrations in excess of 5% but it is common practice in the UK to ensure that levels reach at least 3.5% in smoked salmon and trout, which is sufficient to prevent growth and toxin formation during the normal chilled shelf life of these products. Indeed, guidelines given by the UK Advisory Committee on the Microbiological Safety of Food recommend that products in which non-proteolytic *C. botulinum* may be a hazard and that are stored under anaerobic conditions, i.e. vacuum-packed, should achieve an aqueous salt content of 3.5%.

The smoking process for cold smoked fish products presents a theoretical hazard associated with the potential for growth of proteolytic and non-proteolytic strains of *C. botulinum* through exposure of the fish to elevated temperatures for extended time periods.

Cold smoked fish are produced by smoking fish on racks at temperatures approaching 30°C for many hours (usually overnight, i.e. 12–18 hours). The fish has a salt content c. 3.5% and it is clear that a potential hazard exists associated with the germination and growth of proteolytic and non-proteolytic *C. botulinum* during the smoking process.

Providing salt levels equilibrate to over 3.5% in the aqueous phase of the product, growth of non-proteolytic *C. botulinum* will be minimal, even during the smoking process at 20–30°C. This is demonstrated when using predictive models which usually predict lag times in excess of 1–2 days under smoking conditions. Therefore, providing the organism has not previously exhausted its lag phase during chilled storage, growth of non-proteolytic strains would not occur during this stage. However, smoking temperatures also bring the product into the growth range of proteolytic strains, which are less affected by low salt concentrations than non-proteolytic strains. While the combination of salt level, temperature and time would appear to allow germination and growth, the amount of growth is likely to be minimal during this stage of the process, and as later stages are refrigerated, proteolytic strains should not present a major hazard to the product if smoking is not extended beyond c. 18 hours.

After smoking, the fish is usually chilled to <5°C prior to slicing immediately or after storage for 24–48 hours. Slicing may be done using manually held knives in the traditional processes or using automated hand-held slicers with rotating blades or even using completely automated machinery. The slices are interleaved and layered onto a laminated board to support the thin product and then vacuum packed prior to chilled storage and retail sale.

To produce hot smoked fish, fish is placed on racks in a smoking chamber which is filled with smoke and heated to achieve a product temperature in excess of 68°C for approximately 30 minutes, within an overall process of several hours. This is clearly capable of destroying vegetative pathogens including any germinated spores of *C. botulinum,* but not the spores themselves. Therefore, like any cooked product it is essential that the cooling of the product after hot smoking is rapid, to prevent germination and growth of any contaminating spores.

There are various reports indicating the inhibitory effect of smoke on the germination of *C. botulinum* and, while this may occur on occasion, the mechanisms of inhibition are not clear and the reproducibility of this effect with different wood types is very poor. Therefore, smoke effects cannot be relied on for adding any quantifiable safety factor and it is normally assumed to have limited effect. In hot smoked products, the cooking further reduces moisture content and results in a final product with a relatively high aqueous salt content, in the region of 5%.

Smoked mackerel (kippers) are usually inhibitory to the growth of *C. botulinum* in pack, providing they are maintained under temperatures precluding the growth of proteolytic strains, i.e. < 10°C. Notwithstanding this, there has been a report of a botulism outbreak associated with a hot smoked fish product (Korkeala *et al.*, 1998) where the process appeared to be under fairly good control (40°C for 30 minutes followed by smoking at 60°C for 30 minutes with stepped increases of 70°C for 30 minutes, 75°C for 40 minutes and then reduced to 70°C for 40 minutes, after which it was cooled to 2.5°C in 90 minutes) and where little evidence existed of post-cooking temperature abuse or extended shelf life prior to consumption. The conditions that allowed toxin production to occur and cause the outbreak were not identified.

Final product issues and control

Cooked fish and shellfish

Standard cooked fish and shellfish, in which cooking does not destroy spores of non-proteolytic *C. botulinum*, are a potential hazard with respect to this organism. However, as post-process microbial contaminants causing microbial spoilage restrict the shelf life of these products to less than seven days, such short shelf life should preclude the growth of the organism to toxin-forming levels under conditions of effective temperature control. Like raw fish however, it is possible to extend the shelf life of cooked fish products by storing under gaseous conditions that restrict the growth rate of the spoilage microflora and thereby delay spoilage. However, such conditions rarely inhibit the growth of non-proteolytic *C. botulinum* and so the extended life afforded to these products increases the chances of germination, growth and toxin formation by any spores that may be present in the finished product.

The organism grows readily in cooked fish products and several studies have investigated the potential for cooked fish products to support the organism's growth. Brown and Gaze (1990) inoculated *C. botulinum* type E and non-proteolytic type B spores into sterilised cod homogenates and fillets prior to vacuum packing and cooking to an internal temperature of 70°C for two minutes. The products were then stored at 2°C, 3°C, 5°C, 8°C and 30°C for up to 12 weeks. No significant growth or toxin formation occurred at any temperature over the 12-week period with the exception of 30°C where growth and toxin occurred after 0.7 week by both strains (Tables 4.24 and 4.25). In a further experiment with two non-proteolytic type B strains, Brown *et al.* (1991) inoculated sterilised cod homogenates and fillets using the same approach as described by Brown and Gaze (1990) and incubated the vacuum packs at 2°C, 5°C, 8°C, 15°C, and 30°C for 12 weeks. No toxin was detected after 12 weeks' storage at 2°C but toxin was produced after 10 and 12 weeks' storage at 5°C by one strain of *C. botulinum* type B in cod homogenate, even though the initial levels of the organism had increased by only 2.5 orders of magnitude to log 3.88 cfu/g at week 12 (Table 4.26).

Table 4.24 Growth of non-proteolytic *C. botulinum* type B in cod homogenates, adapted from Brown and Gaze (1990)

Time (weeks)	Log count of *C. botulinum*/g			
	Temperature (°C)			
	3	5	8	30
0.7	1.75	1.94	1.45	6.7*
1	1.38	1.94	1.36	7.66*
4	1.28	0.9	1.38	NT
8	0.6	0.3	0.3	NT
12	0	0.3	0.3	NT

Initial inoculum: 100 spores/g.
*Toxin detected.
NT = not tested.

Table 4.25 Growth of *C. botulinum* type E in cod homogenates, adapted from Brown and Gaze (1990)

Time (weeks)	Log count of *C. botulinum*/g			
	Temperature (°C)			
	3	5	8	30
0.7	3.14	2.79	2.77	7.44*
1	2.84	3	3.1	7.85*
4	2.75	2.6	2.8	NT
8	2.42	2.77	2.58	NT
12	2.47	2.59	2.51	NT

Initial inoculum: 1000 spores/g.
*Toxin detected.
NT = not tested.

Table 4.26   Growth of non-proteolytic *C. botulinum* type B in cod homogenates and fillets, adapted from Brown *et al.* (1991)

Time (weeks)	Log count of *C. botulinum*/g		
	Temperature (°C)		
	5	8	15
0.7	1.2	1.53	2.75
1	NT	NT	2.07*
2	NT	NT	6.89* (6.49*)
4	1.15	0.9	NT
8	NT†	4.04 (4.02*)	NT
12	3.88* (3.07)	5.39* (5.59*)	NT

Initial inoculum: 100 spores/g.
Levels after cooking 3.4×10^2/g.
( ): Data for cod fillets.
* Toxin detected.
† Toxin detected after 10 weeks.
NT = not tested.

At 8°C, growth and toxin production by one type B strain was detected in the cod fillet after 8 weeks and after 12 weeks in the homogenate. In both cases, toxin production was accompanied by 3–4 log cycle increases in numbers. However, with the second type B strain, toxin was detected after only 4 weeks' incubation at 8°C (Table 4.27), although subsequent toxin assays after 8 weeks and 12 weeks were negative. It is interesting that toxin was produced in the absence of any apparent quantifiable increase in levels of the inoculated strain of *C. botulinum* and with levels being reported as log 0.48 cfu/g. This may indicate that toxin accumulated over a long time period, sufficient to result in a positive assay using the mouse bioassay method. Alternatively, it may be that difficulties in culturing the organism failed to detect the real number present. Nevertheless, it provides a warning to those using standard growth predictive mathematical models to predict safety of foods in relation to *C. botulinum* in long shelf life products.

Meng and Genigeorgis (1994) studied the growth of *C. botulinum* type B (non-proteolytic) and E in poached salmon. Inocula of 10^4 spores per 3 g product produced toxin after storage at 4°C, 8°C and 12°C within 60 days, 8 days and 4 days, respectively (Table 4.28).

It is apparent therefore that while different forms of packing may allow extension of cooked fish and shellfish product shelf life if accompanied by either reduced post-process contamination or cooking in pack, non-

Clostridium botulinum

Table 4.27 Growth of non-proteolytic *C. botulinum* type B (strain 2) in cod homogenates and fillets, adapted from Brown *et al.* (1991)

Time (weeks)	Log count of *C. botulinum*/g		
	Temperature (°C)		
	5	8	15
0.7	NT	NT	NT
1	NT	NT	3.52
2	NT	NT	6.43* (7.57*)†
4	NT	0.48*	7.32*
8	NT	0.3	NT
12	0.7	0.3	NT

Initial inoculum: 100 spores/g.
Levels after cooking 2/g.
( ): Data for cod fillets.
*Toxin detected.
†Toxin detected in cod fillets at week 3.
NT = not tested.

proteolytic *C. botulinum* presents a significant hazard if chilled shelf lives are extended beyond 2–3 weeks, where no other control measure is in place.

Smoked fish

A number of surveys have been carried out to examine the incidence of *C. botulinum* in packaged smoked fish. Heinitz and Johnson (1998) recently found no spores of *C. botulinum* in 201 samples of vacuum-packed smoked fish (Table 4.29). Hyytiä *et al.* (1998), however, found the *C. botulinum* type E toxin gene (using a PCR technique) in 11 out of 223 samples of hot smoked fish (including smoked salmon, trout and herring). Levels of the organism isolated using conventional most probable number (MPN) techniques were estimated to be between 30 and 60 spores/kg. *C. botulinum* was also detected in two out of 64 samples of cold smoked trout at 160 spores/kg (MPN technique). Clearly, the incidence of *C.*

Table 4.28 Toxin production by *C. botulinum* type B (non-proteolytic) and E (10⁴ spores per 3 g product) in poached salmon stored at various chill temperatures, adapted from Meng and Genigeorgis (1994)

Temperature (°C)	4	8	12
Time to toxin formation (days)	60	8	4

pH 6.4, brine concentration 0.29% (aqueous salt).

Table 4.29   Incidence of *C. botulinum* in smoked fish

Product	Detection (number positive/ number tested)	Levels (MPN/kg)	Reference
Vacuum-packed smoked fish	0/201	—	Heinitz and Johnson (1998)*
Vacuum-packed cold-smoked rainbow trout	2/64	160	Hyytiä *et al.* (1998)†
Vacuum-packed hot-smoked rainbow trout or salmon	2/50	30	
Vacuum-packed hot-smoked white fish	5/50	40	
Air-packed hot-smoked herring	0/50	—	
Air-packed hot-smoked vendace	3/50	30	
Air-packed hot-smoked river lamprey	1/23	60	

* Using enrichment and isolation procedure (50–100 g samples).
† Using PCR detection of toxin type E gene followed by cultural isolation using MPN (of PCR positive samples).

botulinum will be highly variable, but it is evidently readily detectable in these fishery products.

The greatest hazard to cold-smoked fish products is presented by the growth of *C. botulinum* during the refrigerated shelf life. Levels of aqueous salt in excess of 3.5% would readily restrict growth over the normal shelf life of these products (3–4 weeks). However, these levels do not prevent growth completely and longer lives in excess of five weeks may allow hazardous levels to develop.

Cann and Taylor (1979) studied the growth of *C. botulinum* in brined and/or smoked fish. In naturally contaminated trout, toxin production by type E was inhibited with salt concentrations of 2.5% (aqueous) over 10, 20 and 30 days at 10°C (in smoked and unsmoked fish). At concentrations of 2% aqueous salt, only ungutted fish became toxic, with smoking providing some reduction in the percentage of fish supporting toxin formation (5.6% smoked, brined fish were toxic versus 19% brined, but not smoked fish). Smoked, whole fish (2.5% aqueous salt) inoculated with

100 spores/g of non-proteolytic *C. botulinum* type B were found to be toxic after 30 days' storage at 10°C (90% fish toxic), with type E, 20% were toxic, but none inoculated with type F were toxic. No toxin was produced after 30 days at 10°C when the salt concentration was raised to 3% in the water phase; 3% aqueous salt also prevented toxin formation in whole smoked trout and mackerel inoculated with 100 vegetative cells and spores of a mixture of types B, C, E and F when stored at 10°C for 10, 20 and 30 days. It was also noted that no toxin was produced after storage at 20°C for one day. In smoked, minced trout with salt concentrations of 2%, *C. botulinum* types E and F (100 spores/g) produced toxin after 10 days and 21 days respectively, while type B produced toxin at 10°C after 10 days at 3% aqueous salt concentration, although it was reported that toxin production was erratic.

In recent times it has become common practice for some cold-smoked fish products to be sold to customers via mail order, where advertisements for the products are placed in newspapers and customers purchase direct from the manufacturer. Vacuum-packed products are usually sent via overnight courier without any active chilling mechanisms. The product may be packed in foam packing and, in some cases, may have cool packs added. It usually takes 18–24 hours or more to reach the destination. Clearly, there are significant hazards associated with the sale of vacuum-packed products in this way. Under conditions of elevated temperature there is potential for germination and growth by both non-proteolytic and proteolytic strains of *C. botulinum* and great care must be exercised by the supplier and customer to ensure that abuse temperatures are not reached for excessive times.

C. botulinum must be recognised as one of the most significant hazards to the safety of cooked and/or smoked fish products and appropriate controls must be identified and managed as part of the hazard analysis programme implemented by manufacturers of these products.

CANNED, CURED, SHELF-STABLE MEATS

Canned, cured, shelf-stable meats have one of the most enviable safety records of most food products in relation to outbreaks of botulism. These products have a long sales history with only one reported outbreak associated with canned, cured liver paste believed to have been under-processed following the build-up of spores in the meat (Thatcher *et al.*, 1967 and Tompkin, 1980). Canned, cured meats are considered in most markets to be standard commodity items with little premium attraction and they are consumed by most sectors of the community. Products include canned whole meats such as ham and meat emulsions including luncheon meat. The majority of products include pork as the predominant raw material although other meat species may also be found. The products are sold under ambient storage conditions and have extensive shelf lives allocated to them. They are predominantly consumed without any further processing by the consumer and it is common for canned ham to be used as sandwich fillings.

Description of process

Products in this category, whether whole meat or emulsion-based, are made under similar conditions and common to all is the absence of a process, e.g. a 'botulinum cook', which, in isolation, could prevent any hazard relating to *C. botulinum*. The safety of these products is controlled by a combination of processing conditions and preservation factors which together reduce the number of spores and prevent the growth of *C. botulinum* during product shelf life.

The process of manufacture for emulsion-based products involves comminution of the meat prior to the addition of salt and preservatives such as nitrite by immersion or direct addition (Figure 4.6). Other flavourings such as spice extracts may also be added and mixed together with emulsifying agents including polyphosphates.

Pork is the predominant meat species but other meats such as beef or chicken may also be used, although it is of some importance to recognise that most of the research to determine the safety of canned, cured meat products has been carried out on pork products. The safety of other cured meat species may need to be validated for individual process and preservative conditions.

The meat mixture is filled into cans, the can lid is applied and the closure sealed. The products are then cooked in a retort. Canned, cured meats are

Process Stage	Consideration
Animal husbandry ↓	Health Cleanliness
Animal slaughter and processing ↓	Hygiene Temperature
Meat transport, delivery and storage ↓	Hygiene Temperature Time
Comminuted and reformed bulk meats Bowl chopping and addition of other ingredients (spices, herbs, salt, nitrite, etc.) or	Hygiene Temperature Distribution of preservative factors
Whole joints or pieces of meat Brine injection/tumbling and equilibration of curing salts, e.g. hams and cured meats ↓	Hygiene Temperature Distribution of preservative factors
Fill cans and seal ↓	Can seal integrity Time
Cooking ↓	Temperature Time High/low risk segregation (post-cooking)
Cooling of cans ↓	Hygiene Disinfection of cooling water
Drying of cans ↓	Hygiene
Labelling and storage ↓	Can damage
Distribution ↓	Can damage
Retail ↓	Can damage
Consumer	

Figure 4.6 Process flow diagram and technical considerations for typical canned, cured meats.

processed to achieve temperatures in excess of 100°C and are normally cooked for 30-45 minutes at retort temperatures > 105°C (usually attaining 110-120°C) and then cooled with water or in air prior to drying, labelling and storing. Products may be sold as individual cans (250-450 g) to consumers for direct consumption or they may be manufactured and sold as bulk packs (> 2 kg) to retail outlets where they may be opened and sliced for sale on the delicatessen counter. In most cases, cooked meats intended for delicatessen slicing are subject to lower cooking temperatures (see section on 'Refrigerated, cooked, cured and uncured meat products' in this chapter) as they are stored under chilled conditions. Some bulk canned, cured, shelf-stable meats may however be supplied for further slicing, particularly where conditions of transport or storage do not allow storage under refrigerated conditions, as required for standard chilled cooked meats.

Canned, cured shelf-stable meat products are assigned shelf lives ranging from several months to in excess of two years and this is dictated more by deterioration in the product eating quality rather than by any risk of microbial growth.

Raw material issues and control

The raw materials used in the production of these products are common to a variety of cooked meat products already described in a previous section of this chapter. The predominant meat species is pork, although chicken and beef may also be used. Meat is known to be contaminated with spores of *C. botulinum* and pork, in particular, is often considered to be the meat species most commonly contaminated with the organism. Surveys of the incidence of *C. botulinum* in raw meat show that it may occur at a frequency of 0%–15% and, when present, levels are estimated to be usually quite low, i.e. < 10 spores/kg. Processing conditions of the raw meat can have a significant influence on the presence and levels of bacterial contamination. Contamination of beef is likely to derive from the hide of the animal which is often heavily contaminated with soil and faeces. Preventing dirty animals from entering slaughter houses and the strict application of hygienic slaughter and hide removal practices for cattle are essential for limiting the extent of contamination. Chicken and poultry have rarely been found to be contaminated with *C. botulinum* types A, B and E which may relate to the conditions in which they are reared, i.e. indoors with controlled feed and litter as opposed to outdoors for pigs and cattle where they are exposed to soil. However, chickens can become colonised by *C. botulinum* during rearing and processing from environmental sources and through cross-contamination. The most common type of *C. botulinum* in poultry is type C which appears to be an avian adapted strain (see Table 1.3).

Attempting to achieve absence of *C. botulinum* from any of these raw meat species is not practically possible but every endeavour should be made to limit the extent of contamination by employing good standards of hygiene in rearing and slaughter and effective cleaning procedures for process equipment both during and after production.

In most cases, the contamination of whole cuts of meat will be restricted to the outer surfaces with very little contamination entering the muscle tissue, although the potential for this to occur on occasions through the blood supply during slaughter should not be overlooked. In general, however, whole cuts of meat are likely to harbour the organisms on outer surfaces whereas comminuted meats will have the organisms distributed throughout the whole meat mix by the comminution process. Clearly, this has some important implications with regard to distribution of preservative factors in the two different types of product.

All products are usually made by addition of sodium chloride, sodium nitrite and emulsifying salts such as mono-, di- or polyphosphates of sodium or potassium. These materials are unlikely to contribute additional microbial contamination to the product.

Some comminuted products may have dried herbs and spices added to them during processing and such additions can bring with them high levels of spore-forming bacteria. Herbs and spices are often imported from less well developed countries where growing conditions may be poorly controlled. Once harvested, they may be dried on the ground where they are subject to contamination from soil, dust and insects and then they are bought in bulk by further processors. Some processors may heat-process the materials prior to use but, in most cases, herbs and spices are used with little further processing. As the primary contaminants are likely to be derived from the soil and air as dust, it is common for herbs and spices to carry very high microbial populations, with a significant proportion being spore-forming bacteria. The warm climates from which many such materials are sourced also ensure the predominant microflora is either mesophilic or thermophilic. Therefore, while surveys of the incidence of *C. botulinum* in these materials are limited, it is clear that the use of dried herbs and spices in these products can add high levels of contaminating spore-formers, some of which will undoubtedly be *C. botulinum*.

The safety of shelf-stable, canned, cured meat products is highly dependent on restricting the spore loading of the raw ingredients as the heat process applied only achieves small reductions in *C. botulinum* levels. Guideline levels of spores in raw materials for comminuted meats have been proposed by Hauschild and Simonsen (1985). Average levels of clostridial spores on raw meat in excess of 3 spores/g or mesophilic bacillary spores above 100/g are considered to warrant investigation of the supply and production chain and an increase in the heat process applied. It was also suggested that the mesophilic spore count for spices used should not exceed 5×10^3/g of spice and that the contribution from all other non-meat ingredients (excluding spices) to the contamination load of the final composite raw product mix should collectively be <50 mesophilic spores/g.

Process issues and control

One of the most critical tasks is ensuring that the preservatives controlling *C. botulinum* are present at the correct concentrations and that these concentrations are effectively distributed/equilibrated throughout the meat/meat mix. The preservatives are critical to the safety of these products and systems must be established to demonstrate that the appropriate level and distribution of these preservatives are achieved during the mixing/injection process. Ingoing salt, emulsifier and sodium nitrite levels should be recorded and routine analysis carried out to ensure that the correct final concentrations in the aqueous phase of the product are achieved to maintain safety.

Levels of ingoing salt and nitrite required to inhibit the growth of *C. botulinum* have been the subject of extensive research over many years. While the results of this work can be difficult to interpret due to some conflicting findings, certain combinations of these controlling factors have been shown to give a high degree of assurance with regard to product stability and inhibition of growth of *C. botulinum*. These include levels of ingoing nitrite usually in excess of 100 ppm with the salt content ranging from 3% to 5% in the aqueous phase of the product.

As these products are subject to a subsequent heat processing stage, it could be easy for operatives to regard the need for hygienic practice and effective cleaning to be of cosmetic rather than critical importance. However, as the safety of the product is also dependent on controlling the microbial loading in the ingredients and meat mix it is essential to limit the build-up of microbial contamination at all stages of the process as high levels can compromise the safety of the final product. Poorly handled raw materials and inadequately cleaned equipment, especially bowl choppers, mixers and filling machines can change a situation of an infrequently occurring organism, present at low levels to one in which the organism may cross-contaminate many batches of raw material at high levels. Attention to detail in the operation of frequent equipment cleaning during the production day and a full and effective clean each day are essential to prevent such scenarios from developing.

Clearly, one of the most important stages in the processing of canned, cured meat products is the heat process. Many of the controls relating to canning technology are described in the section 'Low-acid canned foods' later in this chapter, but are equally applicable to these products. Perhaps the most critical stage is the heat process itself.

Canned, cured meats are filled into containers which are lidded and sealed and then cooked for extended periods in a retort. The process is designed to achieve a cook equivalent to F_0 0.5–F_0 2, depending on the raw material and preservative system employed. The retort temperature used to achieve this is usually between 105 and 115°C and it can, therefore, take from 30 minutes to 2 hours or more to achieve the required F_0 process. Attaining the required temperature is critical to the safety of the product and procedures must be in place to ensure this is achieved in all parts of all cans and in all areas of the retort. It is therefore essential to conduct a process validation study to determine all factors that are important for control in achieving the correct cook of the product. For canned, cured meats this can depend on a number of factors including the size of meat pieces used, the density of the raw meat mix, the head space in the can and ingoing temperature of the meat. All of these factors will affect the ability of any given process to achieve the correct temperatures throughout the meat mixture in the can, particularly at the slowest heating point, which is often the centre of the can.

The moisture content of the mixture may affect susceptibility of the spores to heat and therefore the effect of formulations that decrease water activity needs to be carefully understood.

The temperature of the raw meat prior to cooking is a factor of utmost importance in achieving the correct cook of these products, as any heat process designed for raw material at a particular ingoing temperature will be less effective if the raw material temperature is actually lower than this. Monitoring ingoing raw material temperature should be part of the raw material checks conducted on each batch of material being processed. In many cases, however, the meat mix will be tempered or even pre-heated to an even temperature prior to retorting, although this needs to be kept under tight control to ensure even temperature distribution is achieved and also to preclude the growth of microorganisms during any holding stage. Cooking devices can vary significantly in their heat distribution characteristics and it is important to establish that cold spots do not occur in the cooking chamber or, if they do, that the heat process is established based on results from studies where the product has been placed in the coolest part of the retort. In most cases, it is usual for the heat process to be established by determining the process conditions of time and temperature capable of achieving the desired F_0 using the most dense raw material in overfilled cans of lowest temperature in the coolest part of the retort with the fastest cooling cycle. Under such circumstances, a process would have a good safety margin. Process controls should be targeted at monitoring raw material temperature, can filling weight, process time and temperature.

Cooking temperature should be monitored using retort chamber probes connected to a continuous chart recorder and it is essential that any device used for monitoring process efficacy is calibrated at regular intervals to ensure correct readings are being obtained. In addition, it is essential to check heat penetration, using can simulants, and heat distribution in the retort at regular intervals to ensure heat process characteristics are reliably maintained. Although these checks are often considered to be onerous, they should be considered as an important part of a preventive approach to food safety that can be the difference between a continuously safe product and an intermittently unsafe one.

Canned, cured meats may be cooled after processing using water and, as the stability and safety of the products are dependent on a combination of factors including the heat process, it is essential to prevent post-process contamination of the product. This may occur as a result of poor seam integrity and ingress of microorganisms during cooling or as a result of can damage during conveyor transfer, storage, transport and sale. Contamination of cooling water can be minimised by effective decontamination of the water using disinfectants such as chlorine. This has been more extensively reviewed in the section 'Low-acid canned foods' in this chapter.

Final product issues and control

The combinations of ingoing salt and nitrite and the heat process necessary to achieve stability of these products have been reported, although they are based on only limited amounts of work on few meat types. By far the greatest amount of data available concerns the safe combinations for products made from pork and pork emulsions.

It is important to note that safety of these products is dependent on the heat process to reduce levels of contaminating spores together with the formulation to subsequently preclude survivor growth. Safety is therefore reliant on low initial spore loading and increases in the levels of initial contamination can result in survival and growth of *C. botulinum* and toxin production. Therefore, while the safety of these products has been demonstrated over long periods of time, history is no excuse for inadequate processing standards. Indeed, the only recorded outbreak of botulism linked to commercial products of this nature was reported in canned, cured liver paste and this was believed to be due to excessive levels of *C. botulinum* building up in the raw meat and a subsequent inadequate heat process.

Roberts and Ingram (1973) admirably demonstrated the potential for *C. botulinum* to grow in the presence of nitrite, with inhibition being dependent on factors such as nitrite concentration, pH and salt content. In fact, at 35°C incubation, the only combination of factors likely to be used in canned, cured meat products where the strains of *C. botulinum* used were found to be inhibited were pH 6.2, salt concentrations of 5% and 6 % (w/v) and a nitrite level of 100 ppm (Table 4.30). Inhibition also occurred at pH 6.0 with 4% and 5% salt and nitrite concentrations of 200 ppm and 100 ppm, respectively.

Table 4.30 Growth of *C. botulinum* in different concentrations of sodium nitrite and salt, adapted from Roberts and Ingram (1973)

Salt concentration (% w/v)	Maximum nitrite concentration (ppm) permitting the growth of *C. botulinum* at different pH and salt contents		
	pH		
	6.2	6	5.8
0	300 +	300 +	250
1	300	250	250
2	300	250	200
3	250	250	150
4	200	150	100
5	50	50	50
6	50	NG	NG

NG = No growth in the presence of 50 ppm nitrite.
Four strains of *C. botulinum* (types A, B, E and F). Vegetative cell mixture incubated at 35°C.

It is important to note that there may be differences in protection from *C. botulinum* afforded by similar processes used for different meat formulations. For example, luncheon meat (Brühwurst) and canned liver sausage containing 80 ppm nitrite and given an F_0 0.4 cook were shown to achieve control of *C. botulinum* at different water activity levels, 0.97 and 0.96, respectively. This was attributed to protection afforded to *C. botulinum* spores against the effect of the heat process by the higher fat levels in the liver product and to the high iron content which reduces the efficacy of nitrite (Hauschild, 1989)

The combinations of inhibitory factors that have previously been reported to achieve a safe and stable product are shown in Table 4.31. The basis for their safety and stability was extensively reviewed by Hauschild and Simonsen (1985) using data from published challenge test studies together with consumption data and associated lack of illness. Even today, the mechanism whereby the combination of salt, nitrite, pH and heat, control growth of *C. botulinum* and hence product safety is still not fully understood. It is therefore essential that any operator deviating from these established guidelines conducts studies to establish the inherent safety of alternative processes and formulations. It is often possible to extend the margins of safety when the factors and mechanisms contributing to them are fully understood. However, when not fully understood, as is the case with these products, it would be very unwise to go beyond the established limits without sound supporting evidence of safety.

Table 4.31   Combinations of heat process, salt and nitrite employed for the safe manufacture of shelf-stable, canned, cured meats, adapted from Hauschild and Simonsen (1985)

Product type	Heat process (F_0)	Salt concentration (aqueous salt, %)
Meat emulsions, e.g.	1–1.5	3–4
luncheon meats	1	4–4.5
	0.5–1	4.5–5
	0.5	5–5.5
Packed whole muscle meats,	0.3–0.5	3.3
e.g. ham and shoulder	0.2–0.3	3.7
	0.1–0.2	4
Sausages	1.5	2.5

Ingoing sodium nitrite: 150 ppm.

EXTENDED-LIFE DAIRY DESSERTS

There has been an increasing trend to develop and market chilled, perishable products with extended shelf lives. Product development has been marked in the dairy sector where advances in clean-fill technology have enabled the industry to achieve levels of hygienic processing not previously possible without the use of dedicated aseptic fill systems. Products such as mousses, cream-containing desserts and similar products are made in which the major constituent is cream. The products are stored under refrigeration and are allocated shelf lives up to and beyond 15 days. They are eaten by all sectors of the population, with products targeted at children accounting for a large part of this sector. No reported outbreaks of botulism have been associated with these products but their long shelf lives give significant cause for concern in relation to the potential hazard associated with non-proteolytic strains of *C. botulinum*.

Description of process

Dairy desserts such as mousses are made by mixing previously pasteurised cream with a variety of other ingredients including sugar, cocoa powder, stabilisers, thickeners and flavourings in a tank, usually with the application of heat to aid mixing and to dissolve the ingredients (Figure 4.7). The mixture is then subjected to a heat process which can vary significantly from mild pasteurisation of 70–80°C for seconds or minutes to ultra-high temperature (UHT) treatment of 110–140°C for short times (seconds). The product is then either tempered in a buffer tank and hot filled or cooled and filled. In some cases, such as in the manufacture of mousses, air is introduced to create a light bubbly texture prior to filling into containers. These are then sealed and, if not previously subjected to a UHT process, cooled to <5°C prior to storage and distribution to retailers. These products are sold under chilled display and allocated a shelf life up to and in excess of 30 days. Many UHT processed and aseptically filled products are sold under ambient conditions with shelf lives of several months.

Products usually have a neutral pH and the main controlling factor is the high sugar content, which, in combination with the low moisture content, create a low water activity, although the presence of oxygen and the high redox potential of aerated mixtures may also play a role in restricting growth of *C. botulinum*.

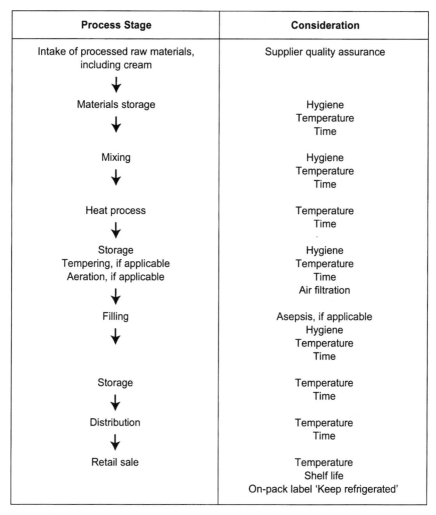

Process Stage	Consideration
Intake of processed raw materials, including cream	Supplier quality assurance
Materials storage	Hygiene Temperature Time
Mixing	Hygiene Temperature Time
Heat process	Temperature Time
Storage Tempering, if applicable Aeration, if applicable	Hygiene Temperature Time Air filtration
Filling	Asepsis, if applicable Hygiene Temperature Time
Storage	Temperature Time
Distribution	Temperature Time
Retail sale	Temperature Shelf life On-pack label 'Keep refrigerated'

Figure 4.7   Process flow diagram and technical considerations for typical dairy desserts.

Raw material issues and control

A variety of raw materials are used in the manufacture of dairy desserts, the most common being cream. Like any material derived from milk, cream will occasionally be contaminated with spores of *C. botulinum*. The frequency will vary but it is generally noted that cream tends to have a slightly higher microbial loading than milk as the process of separating whole milk into cream and semi-skimmed or skimmed milk tends to concentrate the bacteria in the cream phase. Few surveys have been published in this area but it can be reasonably assumed that the frequency of contamination with *C. botulinum* will be equivalent to or slightly higher than whole milk, which, if present, is likely to occur at levels of one cell/litre or less (Collins-Thompson and Wood, 1993).

The organism is most likely to enter the milk supply via the same route as other contaminating microorganisms, i.e. during milking. Contamination from the udder is the most likely source with the organism present on the outside of the udder in the general soiling often found on udders prior to milking. Such soiling will consist of faeces, soil and straw and these are well known sources of *C. botulinum*. The operation of hygienic milking practices including cleaning the udder and teats before and after milking together with the use of teat disinfectants will clearly reduce the microbial loading entering the milk supply but cannot exclude it completely.

Contamination may also occur from poorly cleaned or maintained milking equipment which can provide foci for colonisation by general bacterial populations. Operation of effective cleaning regimes is important to prevent the build-up of contaminants although the likelihood of *C. botulinum* colonising these sites is probably quite low due to competition from other contaminants such as pseudomonads or Enterobacteriaceae.

Other raw materials used in the manufacture of dairy desserts may also contribute to the microbial loading of the raw material mixture, with dried ingredients being the most likely sources. Highly refined material such as sugar has low microbial loads and is unlikely to be a source of contamination but other ingredients such as natural flavourings and cocoa powder may carry much higher microbial loads and the risk of the presence of *C. botulinum* should always be considered in a hazard analysis of the product's ingredients and manufacturing process.

It is good practice to monitor the spore loading of relevant raw ingredients at defined frequencies against an agreed supply specification as an increase may give some indication of poorer quality raw materials that may have been subject to lower standards of hygiene or process control. Although general aerobic spore counts do not in themselves indicate that *C. botulinum* is present, it is reasonable to use them as an indicator of conditions employed at the supply site which could in turn indicate a greater potential to allow the entry and proliferation of spore-bearing bacteria. If an anaerobic mesophilic spore count is used to monitor incoming raw materials, this may provide a more appropriate marker for any increasing potential for *C. botulinum* to be present, although there is unlikely to be a direct relationship between the levels of general spore counts and the levels of *C. botulinum*. Any increase in levels of spores found as part of a routine monitoring programme requires further investigation in conjunction with the raw material supplier.

In general, however, it is not possible to exclude totally the organism from raw materials used for the manufacture of dairy desserts and so the hazard must be anticipated to exist and needs to be controlled in subsequent stages of the process and by the product formulation.

The primary hazard to dairy desserts which do not receive a 'botulinum cook' or which are subject to post-heating contamination is the non-proteolytic strains of *C. botulinum* as products in this category are intended to be stored under refrigeration conditions. The relative frequency with which these types occur in milk as opposed to proteolytic strains is not known but, as contamination is likely to be originally derived from soil or faecal material adhering to the cow's udder and from the dairy environment, it is likely that types A and/or B will be more prevalent in milk and cream. Thus, non-proteolytic *C. botulinum* type B would represent the greatest hazard for consideration in relation to these products.

It is important to ensure that any perishable raw ingredient, particularly the cream, is stored under controlled refrigerated conditions with a shelf life precluding the growth of *C. botulinum*. For dairy desserts which do not receive any heating during processing, i.e. cold mixing only, the shelf life of the raw ingredient should be taken into account when setting the shelf life of the finished product. This is particularly important if it is deemed capable of supporting the growth of *C. botulinum* as the risk of growth will be cumulative, based on the combined life of the raw ingredient, e.g., cream and the subsequent life of the final product.

With products receiving a heat process capable of destroying vegetative cells, this may be less important because any spores which germinate in the raw ingredient will be destroyed. However, as any heat process is designed to achieve a certain microbial reduction, it is always important to limit the loading of contaminants in the first instance by operating high standards of raw material control.

Process issues and control

Most products of this nature such as mousses or desserts are made up in pre-cook tanks by mixing the selected raw ingredients into the cream or milk. It is important to ensure that all equipment and the environments in which such materials are processed are cleaned effectively and maintained hygienically. A heat process must not be used or regarded in any way as cleaning up an inadequately controlled raw material or a raw material extensively contaminated by poorly maintained and unhygienic equipment or through temperature abuse.

The formulation of the mix is often the factor that controls the safety of non-UHT, extended-life dairy desserts. In these situations the ingredients contributing to safety are clearly critical and must be added at the correct concentrations and monitored appropriately. For example, if water activity is the controlling factor, then the level of humectant such as sugar or even stabiliser is critical to control as this, combined with the water content, will affect the water activity of the product. These should therefore be subject to stringent process monitoring and control which will involve accurate weighing of ingredients and full process records of the amounts added against a documented ingredient sheet. It should also include measurement of the water activity and other parameters important in controlling microbial growth such as moisture and sugar content (°Brix).

Where control is exerted by maintenance of low pH then this should be subject to routine monitoring and strict limits set on the process. The nature of the acidulant may also be an important consideration in the control of *C. botulinum* in pH-controlled dairy desserts.

One of the most critical stages of these manufacturing processes is likely to be the heating stage. Many products receive a high-temperature, short-time pasteurisation process with temperatures usually in excess of 80°C. Where the process is intended to destroy the spores of non-proteolytic strains of *C. botulinum*, it should reach at least 90°C for 10 minutes. Although there is some debate about the level of protection afforded by such processing, it is generally considered to deliver a safe product with regard to the hazard from non-proteolytic strains of *C. botulinum* for products destined for controlled storage under chill temperatures (<8°C).

Control of time and temperature of the heating process is essential to ensure that processing achieves the correct lethality. Care must be taken with dairy products to account for thermal protection that may be afforded to the organism in low water activity, high fat content dairy desserts. However, the water activities of products affording significant protection against thermal destruction are probably so low that the organisms, if present, could not grow anyway, i.e. water activity 0.94 or less. This, however, may be a more important consideration for UHT, low water activity products which are sold under ambient display conditions in which the water activity could allow proteolytic strains to grow while also providing some protection against destruction by the heat process. It is well established that low water activity provides a protective effect for microorganisms in heat processes and although no specific guidance is available, this effect should be recognised for low water activity dairy desserts where a 'botulinum cook' is being applied. In some circumstances, it may be necessary to conduct challenge tests and lethality studies with strains of *C. botulinum* in the actual product (in research facilities) to determine the processing required to achieve reliable destruction.

Some dairy dessert products may be UHT processed at temperatures of 135°C for several seconds. Under these conditions the destruction of non-proteolytic and proteolytic *C. botulinum* spores is achieved. At a temperature of 131°C, a heat process of 18 seconds would give an F_0 3 (assuming a *z* value of 10) and at 141°C only 1.8 seconds would be required to achieve an F_0 3.

Collins-Thompson and Wood (1993) reported destruction of *C. botulinum* types A and B (1000 spores/ml) in milk after processing at 125°C for 5 seconds. Processing at 122°C for 4 seconds and 116°C for 3 seconds both allowed survival of spores which subsequently grew in processed milk.

With all processes involving a heating stage it is clearly essential that the appropriate controls are in place to ensure the times and temperatures necessary for destruction of a specified hazard are achieved. This will necessitate process validation studies to determine the minimum process times and temperatures required to achieve the full process lethality under worst case conditions, taking account of the coldest raw material entering the coolest part of the heating equipment for the shortest duration and it must also take account of the product characteristics, e.g. water activity, viscosity, etc.

Such validation studies need to be supplemented with process controls and monitoring of all critical stages to ensure the process is consistently achieved and, if not, appropriate process failure indicators are in place such as alarms or divert valves.

If a 'botulinum cook' (whether for non-proteolytic or proteolytic strains) is applied to the product and this represents the sole controlling factor, then it is clearly critical to ensure that post-process contamination of the product does not occur. Contamination of the product may occur due to poor cleaning or sanitisation of the post-heating pipework, holding tank and filling heads or even packaging. Alternatively, contamination may arise from the environment due to airborne contamination entering non-aseptic post-processing areas. Products of this nature should ideally be filled into packs using aseptic filling equipment which limits any opportunity for post-process contamination. Alternatively, the product could be heated in pack.

Many extended-life, refrigerated dairy desserts are, however, packed under non-aseptic conditions, under what is often referred to as 'clean fill' conditions. This usually consists of equipment and filling heads that are subject to high-temperature sanitisation and where the product is filled into containers under a filtered air 'curtain'; this process falls short of aseptic filling conditions. Clearly, this is an improvement over filling lines where no air 'curtain' is present and it should significantly limit any post-process contamination. The extent to which this would reduce the risk of spores of *C. botulinum* occurring as a post-process contaminant from the air or the equipment is a matter for individual companies to assess. As an additional measure, some processors also fill at high temperatures (> 70°C) which will preclude the build-up of vegetative contaminants, but the effect on any spores present will be minimal. Some companies may consider the controls in 'clean fill' operations to provide an acceptable level of risk associated with the presence of non-proteolytic strains of *C. botulinum* in the product and therefore assign long chilled shelf lives.

Indeed, it is more likely that if products were to be post-process contaminated, this would generally be by spoilage microorganisms which would be expected to cause microbial deterioration of the product well before the growth of *C. botulinum*, if present. However, many companies would consider the risk associated with post-process contamination in these types of non-aseptic filling machines to be too great with regard to *C. botulinum* and therefore would require that either the product was aseptically filled or that other controlling factors were present and capable of restricting the growth of the organism. Such decisions must rightly rest with individual companies and they must decide the level of risk that they are prepared to take. Perhaps the adoption of quantitative risk assessment approaches in the future may allow quantification of the level of risk involved, allowing more informed decisions about whether specific practices are likely to compromise public health or not.

Products subject to UHT processes and aseptic packing also require the employment of strict hygiene and process controls. Equipment used for such purposes requires rigorous maintenance and effective cleaning procedures to be employed. In addition, decontamination procedures for the packaging using peroxide or ultra-violet (UV) light systems must ensure that contamination entering from these sources is minimal. It is usual to carry out some packaging checks to monitor the general microbial contamination levels on/in supplied packaging to verify the hygienic status of the packaging against an agreed specification and to monitor the practices employed at the packaging supplier. Commonly expected standards for food packaging would be 'absence' of microorganisms from swabs of the packaging, i.e. none detected per 25 cm^2 or < 1/ml in rinses, with only the occasional detection of contaminants allowed.

The development of in-line, semi-continuous monitoring systems to assess cleaning efficacy using ATP bioluminescence assays of rinse water from in-place cleaning systems is a welcome development for helping to optimise cleaning and sanitisation routines. Removal of product debris can be readily monitored and cleaning regimes optimised with the generation of real-time results.

Final product issues and control

There are few published reports on the incidence of *C. botulinum* in dairy desserts and in particular of non-proteolytic strains. However, *C. botulinum* spores were found in a variety of fermented dairy products in a survey by Franciosa *et al.* (1999). They found no *C. botulinum* spores in ricotta cheese (33 samples), raw milk (35 samples), pasteurised milk (13 samples), clotted milk (3 samples), butter (2 samples) and cream (62 samples) but *C. botulinum* type A was found in mozzarella cheese (5 out of 30 samples), soft cheese (1 out of 81 samples) and processed cheese (1 out of 1 sample).

Dairy desserts usually have a neutral pH and high moisture content and control of non-proteolytic *C. botulinum* is only achievable by three practical means: destruction of the organism by heat processing at 90°C for 10 minutes followed by aseptic handling practices, inhibition of growth using sugar and/or low moisture to reduce water activity to 0.97 or less or by limiting the shelf life. The very nature of extended shelf life dairy desserts makes the latter option not possible and most products have shelf lives well in excess of 10 days and some as long as 30–40 days. UHT, aseptically filled products are assigned much longer shelf lives, often at ambient temperature, of several months.

Products which may be subject to contamination by non-proteolytic *C. botulinum*, either as process survivors or as post-process contaminants, with shelf lives in excess of 10 days should be considered to represent a potential hazard associated with its growth under refrigerated conditions (unless control of temperature would ensure that $<3°C$ was achieved throughout the shelf life).

Other factors have also been postulated to inhibit *C. botulinum* growth in dairy products. Factors such as autoreductive properties, degree of processing, redox potential and even the age of the product have been reviewed by Collins-Thompson and Wood (1993) but many of the reports are contradictory and attempting control by reliance on these factors is unreliable.

For example, although it is possible that the presence of oxygen or high redox potentials, e.g. above $+250\,mV$, in many products may be capable of restricting growth of *C. botulinum*, it is also evident that a high redox potential in many products may allow growth initiation from spores, i.e. up to $+200\,mV$ (Hauschild, 1989). It is considered by the authors to be unwise to rely on the presence of oxygen and high positive redox potential in the product as the sole means of controlling *C. botulinum*. While they undoubtedly play a role, these parameters may not be consistent throughout the product and the levels may also vary throughout the shelf life of the product making reliance on them unsafe. In some cases, it may be necessary to support/confirm these assumptions by the use of challenge tests to determine the level of growth of the organism in the actual product under consideration. Even then, it is possible that significant variation could occur from production to production or between production batches, and reliance on these factors is not advocated.

The best way to achieve control of *C. botulinum* in these products is by the use of inhibitory levels of pH or water activity. Although a pH of 5.0 or less would ensure safety with respect to non-proteolytic strains, unless the dairy dessert is a yoghurt, most would not tolerate such levels of acidity without significant adverse effects on flavour. Therefore, control is usually achieved by low water activity alone. It is generally accepted that water activities of 0.97 or below would prevent growth of non-proteolytic strains, but this has historically been derived from work using sodium chloride as the humectant. These strains have been shown to grow to water activity levels as low as 0.94 where glycerol was used as the humectant.

Predictive mathematical growth models are readily available for non-proteolytic *C. botulinum* and this is a very useful and welcome development. This has allowed product formulation changes to be assessed for the effect on growth of the organism and, in situations where product safety may appear to be questionable, it is useful to be able to draw on predictive model information. However, predictive models are, rightly, fail-safe but the extent to which this is reflective of the real product is highly variable as many factors affecting the control of pathogens in complex foods are poorly understood and not included in predictive models. These models should therefore be used with caution and only by those trained and experienced in their benefits and limitations.

However, in general, predictive growth models can be a useful additional source of data relating to pathogen control. For example, the Food MicroModel (1999) predicts growth of non-proteolytic *C. botulinum* from initial levels of one in a product of neutral pH (7.0) and water activity of 0.99 under chilled conditions (5°C) to 1000 within 17 days. Reducing the water activity to 0.98 would extend this period to 44 days.

Clearly, this does not give information on time to toxin development and indeed, is based on a_w values where salt was used as the humectant, and these may be important considerations when setting safe shelf lives. It does, however, allow an assessment to be made about the potential for growth and the levels that might be reached after a certain period of shelf life.

Dairy products clearly support growth and toxin production by *C. botulinum*, although the literature indicates slower times to toxin production than might be predicted. Some work has been reported on the growth of non-proteolytic strains of *C. botulinum* in milk under refrigeration temperatures. Read *et al.* (1970) inoculated six strains of *C. botulinum* type E into sterile whole milk and assessed toxin production at 4.4°C, 7.2°C, 10°C, 15°C, 20°C and 30°C. Toxin was produced at all temperatures with the exception of 4.4°C (Table 4.32). Not surprisingly, toxin was produced rapidly (2–4 days) at 30°C, with the time to toxin production increasing at lower temperatures. Glass *et al.* (1999) studied the effect of carbon dioxide (0 mM, 9.1 mM and 18.2 mM) in HTST pasteurised milk (2% fat) on growth and toxin production by a 10 strain mixture of proteolytic and non-proteolytic *C. botulinum* (types A, B and E) inoculated at 10^1–10^2 spores/ml and stored at 6.1°C or 21°C for 60 days and 6 days, respectively. No toxin was produced in any of the milk stored at 6.1°C for 60 days and the milk had a normal odour and pH (c. 6.5) up to day 21. At 21°C the milk curdled after 2 days and was putrid after 4 days. Toxin was produced in all spoiled samples after 6 days irrespective of the treatment and after 4 days in the milk treated with 9.1 mM CO_2.

Table 4.32 Toxin production by *C. botulinum* type E in sterile whole milk, adapted from Read *et al.* (1970)

Temperature (°C)	4.4	7.2	10	15	20	30
Time to toxin (days)	No toxin	70	21-56	14-35	3-28	2-4
Number of strains producing toxin/number examined	0/6	1/6	4/6	6/6	6/6	6/6

Even if the formulation of the product is designed to control growth of non-proteolytic *C. botulinum*, it is essential to recognise that such conditions may not preclude the growth of proteolytic strains if the product temperature is not maintained below 10°C. Indeed, the outbreak of botulism associated with mascarpone is testimony to this fact (high pH, high water activity, long shelf life, chilled dairy product). It is therefore essential for these products to be clearly labelled to 'Keep refrigerated'. As the shelf life is also important, it is relevant for products to be clearly marked with a durability indicator that informs the purchaser of the date beyond which the product should not be consumed. This is usually achieved using statements such as 'Use by' or 'Consume within' a specified date on the packaging.

These products have an excellent safety record with regard to botulism but they do have the potential to cause incidents and outbreaks if not controlled effectively.

AMBIENT-STABLE, EXTENDED SHELF LIFE PASTA AND
BREAD PRODUCTS

Bread and pasta have been a component of most people's diet for centuries. Although manufactured from a variety of different ingredients, the processes employed for their manufacture have remained fundamentally unchanged through time. Common throughout all of this time has been the natural restriction of shelf life of bread and fresh pasta by the growth of contaminating moulds. Strategies to extend shelf life have principally employed drying, particularly in the case of pasta, and the use of preservatives, in the case of breads. Product conditions or formulations employed to restrict the growth of spoilage microorganisms, particularly the moulds, have generally resulted in restriction of the growth of some potentially harmful microorganisms as well.

In recent years, the advent of new packaging technologies, e.g. modified atmosphere packing or vacuum-packing, has benefited bread and fresh (higher moisture) pasta products by allowing much longer shelf lives with respect to both microbial spoilage and product eating quality. Many bakery products such as part-baked loaves and rolls, pitta bread and crumpets have been marketed as premium lines with the added benefit of long, ambient-temperature shelf stability. Likewise, 'fresh' pasta (some incorporating cooked fillings) that otherwise would have spoiled has been moved out of the chill cabinet and onto ambient display. These products have opened up a large new market sector which is enjoyed by many sections of the population but they have also brought with them some potential hazards, particularly those associated with *C. botulinum*.

Ambient-stable, extended shelf life bread and pasta products are among an increasing range of products that have benefited from new technological advances allowing extension of shelf life. However, they do not have a long sales history and the hazards associated with them are generally poorly researched.

Description of process

The manufacturing process for bread and pasta differs significantly (Figure 4.8), although the major components of wheat and water are the same, albeit, in the case of wheat, of very different varieties.

Bread, in its simplest form, is manufactured by mixing flour, water, salt, lard or oil and yeast to form a dough. Preservatives, usually vinegar, may be added, although potassium sorbate may also be used to inhibit mould

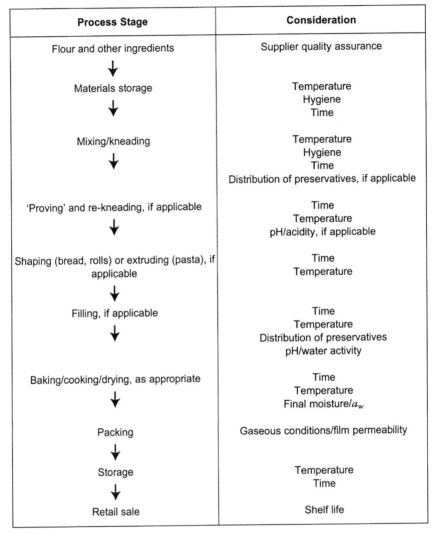

Process Stage	Consideration
Flour and other ingredients	Supplier quality assurance
Materials storage	Temperature Hygiene Time
Mixing/kneading	Temperature Hygiene Time Distribution of preservatives, if applicable
'Proving' and re-kneading, if applicable	Time Temperature pH/acidity, if applicable
Shaping (bread, rolls) or extruding (pasta), if applicable	Time Temperature
Filling, if applicable	Time Temperature Distribution of preservatives pH/water activity
Baking/cooking/drying, as appropriate	Time Temperature Final moisture/a_w
Packing	Gaseous conditions/film permeability
Storage	Temperature Time
Retail sale	Shelf life

Figure 4.8 Process flow diagram and technical considerations for typical pasta and bread products.

growth. The dough is kneaded prior to distribution into baking tins and then left at ambient temperature ('proved') for periods ranging from an hour to several hours depending on the product. The yeast fermentation during 'proving' results in the production of ethanol and organic acids which may contribute flavour and carbon dioxide which causes the dough to rise. The dough is then placed in an oven at temperatures in excess of 200°C and baked to achieve a soft texture in the centre and a crust on the outside. The bread crust and body texture are very dependent on the baking conditions and, in most processes, the centre of the bread reaches close to 100°C for several minutes.

Some bread is manufactured using 'mother' doughs, where an aged batter/dough is produced by extended fermentation usually over 12-18 hours. This results in the development of a strong flavour due to the growth of both yeasts and lactic acid bacteria. A small proportion of this is then added to new dough for production of the bread.

Some bakery products such as crumpets do not receive a traditional baking process. These products are referred to as 'hot-plate' goods because they are cooked on a pre-heated hot metallic surface or plate. Hot-plate goods are usually made from a batter mix consisting of wheat flour, water, salt and vinegar and some also have yeasts added and then allowed a very short fermentation time.

The key feature of all of these bakery type products is that the baking/cooking stage does not reach temperatures capable of destroying strains of proteolytic *C. botulinum*.

Pasta products are manufactured similarly by mixing water with flour (using durum wheat) together with salt in a large continuous mixer to achieve a fairly low moisture dough (30-40% moisture). The pasta may also have liquid egg added to it at the mixing bowl stage. The formed dough passes via an auger to an extruder where it is forced under high pressures through a metal 'die' and exits as a defined shape, e.g. spaghetti or spirals. Alternatively, the pasta may be rolled into sheets and shapes cut out as required.

If the pasta is to be filled, a pre-cooked filling is produced by mixing the ingredients, commonly consisting of a tomato base with meat or cheese together with salt, flavourings, herbs and spices and then boiling the mixture for a short period. The chilled mixture is filled into the centre of a pre-cut flat pasta shape which is then sealed by either folding the pasta over the mix or by placing a second pasta sheet over the first, on which the filling has been deposited, and sealing them together. These filled pasta parcels may then be pasteurised by passing them through a steam cooker to reach temperatures above 90°C.

Filled and unfilled pasta products are usually dried to reduce their moisture level and the drying process can vary significantly in duration depending on the required final moisture level. Temperatures usually reach 60-70°C and are progressively lowered during the drying process.

Like bakery products, none of the stages employed during the manufacture of pasta products are capable of destroying the spores of proteolytic strains of *C. botulinum*.

All products after processing are tempered in air at ambient temperature and then packed. Extended-life products are packed in modified atmospheres or by vacuum-packing. Gas mixtures are usually based on carbon dioxide and nitrogen with the intention being to exclude as much oxygen as possible to achieve long mould-free shelf lives.

Shelf lives can vary considerably, with higher moisture products like hotplate goods being allocated lives of two weeks or more while drier breads and pasta may be given shelf lives of several months.

Most bakery products are consumed with little or no further cooking although some may be warmed in the oven or toasted. A few part-baked products will be baked by the customer, but centre temperatures rarely achieve more than 70–80°C as the re-baking required is merely to brown the crust.

Pasta products are all cooked prior to consumption due to the need to cook the usually raw pasta component in boiling water.

Raw material issues and control

Bread and pasta products have one major ingredient in common and that is flour. Being derived from different wheat varieties, the nature of flour differs from product to product, but the inherent hazard associated with the potential presence of spores of proteolytic *C. botulinum* remains the same in all wheat and all flour types.

Wheat is a commodity crop grown throughout the world and it is subject to significant microbial contamination during growing, harvesting and processing into flour. The microflora of wheat will primarily be derived from the soil, air and water and will include a large variety of microorganisms including yeasts, moulds and bacteria. Contamination will primarily be on the outer parts of the plant although colonisation of the seed may also occur. The incidence and level of contamination of wheat with *C. botulinum* will vary depending on the levels and types in the environment in which the wheat is grown and harvested but, in general, will consist of proteolytic strains of types A and B. Levels of contamination will vary significantly depending on subsequent processing practices but they are not likely to be any higher than those found in the general environment. Contamination will be exacerbated during harvesting due to soil and dust and will also be affected by the storage and milling processes further down the line.

Removal of husks will reduce microbial loading as much of the contamination will be associated with the external surfaces but it is clear that contamination of wheat and subsequently flour with spores of *C. botulinum* is inevitable. Few surveys have been conducted for *C. botulinum* in field grains but the general microflora will include high levels, up to 10^5 spores/g, of spore-bearing bacteria (International Commission on Microbiological Specifications for Foods, 1998) and, on occasion, will include this organism.

Few other ingredients of bread are likely to contribute significantly to the level of bacterial spore-loading when compared to flour, but it is important not to overlook other raw ingredients as potential sources of spores and if intervention measures can be applied such as washing, they should be considered, particularly for bread containing ingredients such as dried fruit or herbs.

Other ingredients are perhaps of more significance to pasta which is to be filled. Such ingredients include an assortment of meats, dairy products and vegetables in fresh, dried or pre-cooked form. It is important to ensure that all raw materials are included in suitable raw ingredient quality assurance programmes and to monitor the microbial loading against agreed specifications to ensure that supplying sites consistently meet quality requirements.

Processing issues and control

The primary hazard with respect to growth of *C. botulinum* does not generally arise with regard to the bread process itself and is usually only a potential concern in the final product. The fermentation process employed for most bread products usually generates a highly competitive microbial environment which is dominated by high levels of yeasts generating metabolites that serve to inhibit the growth of many contaminating organisms. Doughs generate both acids and alcohols and even at the warm temperatures employed during 'proving', it is generally considered unlikely that growth of *C. botulinum* would present a hazard in a normal active dough fermentation. Clearly however, control of the fermentation or indeed, the times and temperatures for those processes where fermentation may not be employed, is important in preventing the growth of contaminating microflora including *C. botulinum*; this may be especially the case in extended fermentation processes. In these cases, it is important to monitor the efficacy of the fermentation process using simple pH or acidity measurements.

Bread is baked in an oven at high temperatures (> 200°C) for varying times, usually dictated by the degree of crust required. Temperatures in the centre rarely exceed 100°C for many minutes and are completely inadequate to achieve any reduction in the numbers of contaminating spores of proteolytic *C. botulinum* types A and B. This is true for most products in this group, including hot-plate goods. Little control is therefore exerted at this stage of the process in relation to this hazard and it must be assumed that any spores of types A and B present in flour and other ingredients will remain in the final product.

Although still unlikely to represent a significant hazard, the process employed for the manufacture of fresh filled pasta may represent more of a concern in relation to the potential for growth of *C. botulinum*. Fillings used in pasta products are not generally ambient-stable and appropriate controls need to be in place to ensure that these are not left at ambient temperatures for excessive times. Many of the fillings are prepared in bulk and may be stored for several days under refrigeration prior to use in filling pasta. Filling takes place at ambient temperature and, as some pasta manufacturers operate on a continuous production basis, running the pasta line for several days, control of the filling is essential. Usually, the pasta component has sufficiently low water activity to preclude growth of most bacteriological hazards but the filling is not usually so stable. Even the tomato-based fillings usually have pH values above 5.5 and water activities above 0.95. Control of this operation can usually be achieved by manufacturing bulk quantities of filling, sufficient for just one production run over a short time period, and cleaning down the filling line at regular intervals.

Like bread, pasta is usually exposed to some form of heating prior to packing. This may be in the form of a pasteurisation or a drying process. Temperatures used are significantly less than those employed for baking and usually range from 50°C to 100°C. Heating is usually applied to dry the pasta to the desired moisture levels, although some products are pasteurised to prevent mould spoilage. Like bread, however, the temperatures used will have no effect on the proteolytic strains of *C. botulinum* and these heat processes cannot be considered to control the organism.

Final product issues and control

Extended shelf life bread products are usually packed in gas impermeable packaging and gas flushed with carbon dioxide and nitrogen. Some bread products such as pittas may be vacuum-packed but most would collapse under such vacuum. Clearly, to achieve significant shelf life, these products must be packed to attain and maintain a low oxygen atmosphere within the pack. Levels achieved during gas packing are usually < 1% although this can be difficult to achieve for larger breads where oxygen may be retained in the loaf.

Bread products vary considerably in pH and water activity and this may also vary between the surface and the centre of the product, the crust usually having a lower initial water activity than the moister centre. Bread has a pH of 6–6.5 and a water activity between 0.95 and 0.97. This is in the range of growth for *C. botulinum* and gaseous conditions capable of restricting growth of moulds in the packed product would generally be a suitable gaseous environment for the growth of *C. botulinum.* These products have historically been allocated shelf lives of several weeks. Drier products such as pittas may have water activities close to or below 0.94 and in many cases, therefore, preclude the growth of *C. botulinum* altogether. These have been allocated much longer shelf lives of several months. At the other extreme, higher moisture products, like many hot-plate goods such as crumpets, usually have water activity values between 0.96 and 0.98 and, under vacuum or modified atmosphere packaging conditions, have had allocated shelf lives of two weeks or more. In many cases, the product may not just be susceptible to the growth of *C. botulinum* but also to other spoilage bacteria including bacilli, and shelf life may be restricted by the potential for these organisms to grow.

Although bread products have not been implicated in outbreaks of botulism, their potential to support the growth of *C. botulinum* has been recognised over the years. The most significant work in this area was conducted by Denny *et al.* (1969). In one experiment, canned breads (chocolate and nut bread and date and nut bread) were inoculated with spores of *C. botulinum* types A and B at levels of 3×10^5/can. The bread was steamed at 208°F (97.8°C) for 60 minutes with the can seam clinched to allow passage of gas as the bread rose. The seam was then completed and then the cans were further steamed at 212°F (100°C) for 70 minutes (60 minutes for date and nut bread). The cans were air-cooled for three days (four days for date and nut bread) and then stored at 85°F (29.4°C) for up to two years and then tested for toxin production. None of the 123 cans of chocolate and nut bread inoculated with *C. botulinum* were positive for toxin production after two years. This was also found to be the case for the 125 cans of date and nut bread. This was attributed to the formulations of the products which resulted in low water activities (Table 4.33).

A further experiment was conducted by Denny *et al.* (1969) to determine the relationship between the water activity of canned breads and the ability to support/inhibit toxin production by *C. botulinum* types A and B. Commercially produced canned bread (corn bread, brown bread, brown bread with raisins, spice loaf, chocolate nut loaf, fruit nut loaf and orange nut loaf) were inoculated post-processing with spores of *C. botulinum* at a level of 2×10^4/can. Cans were incubated at 85°F (29.4°C) for up to one year and tested for botulinum toxin. Although individual formulations are not reported, toxin production was recorded in canned bread with a_w 0.955 or above. No toxin was detected in canned breads with a_w at or below 0.950 (Table 4.34). It is reported that breads of high a_w allowed rapid growth and toxin production by *C. botulinum* (corn bread with an a_w of 0.972 tested positive for toxin after only 16 days). It should also be noted that while the a_w was apparently high, the moisture content was only 33.76%, demonstrating the importance of control based on water activity rather than moisture content in these products.

Aramouni *et al.* (1994) assessed the stability of home-style, canned quick breads – designed to be baked in the home and then stored prior to consumption. They inoculated dough with spores of *C. sporogenes* (10^4/g) and subjected the doughs to several different baking regimes (177°C for 30 min, 40 min and 50 min, 191°C for 45 min, 50 min and 55 min, and 204°C for 40 min, 45 min and 50 min). Spores were detected after each baking regime although little growth was observed after 90 days' storage either at ambient temperature or at 35°C. The post-baking water activities ranged from 0.93 to 0.95 with pH values of 7.6–7.9.

Daifas *et al.* (1999) studied the growth and toxigenesis of proteolytic *C. botulinum* types A and B in high water activity, 'English-style' crumpets. Spores of four type A and five type B strains were composited and used to inoculate crumpets before and after cooking. Crumpets were cooked for approximately nine minutes on an oiled griddle, pre-heated to 204°C. After cooking they were stored at 25°C under various atmospheric conditions; air, air + oxygen absorbent and CO_2/N_2 mixture (60 : 40 ratio). Levels of inoculum in both studies were 5×10^2 spores/g added either pre- or post-cooking. The temperatures achieved in the crumpet during the cooking approached, but did not exceed, 97°C, with a calculated equivalent process lethality at 121°C of 0.03 minute.

Table 4.33 Growth and toxin production by *C. botulinum* types A and B in canned breads stored at 85°F (29.4°C), adapted from Denny *et al.* (1969)

Product	Time of test	Moisture (%)	pH	Total sugar (%)	Invert sugar (%)	Water activity	Toxin production
Chocolate and nut	Immediate	37.3	6.6	25.8	11.3	–	–
	2 years	39.0–40.5	6.1–6.6	26.6–29.5	10.2–11.1	0.922–0.929	0/123
Date and nut	Immediate	45.8	6.7	20.7	19.3	–	–
	2 years	46.5–50.0	6.1–6.4	20.6–22.9	16.5–17.7	0.938–0.948	0/125

Inoculum 3×10^5/can.
Bread processed at 208°F (97.8°C) for 60 minutes and then 212°F (100°C) for 60–70 minutes and then stored at 85°F (29.4°C) for two years.

Table 4.34 Growth and toxin production by *C. botulinum* types A and B inoculated post-processing at 2×10^4 spores/can into seven varieties of canned bread and stored at 85°F (29.4°C) for one year, adapted from Denny *et al.* (1969)

Water activity	Number of cans	Toxic cans
0.955–0.977	56	28
0.855–0.950	73	0

All products stored under each condition remained organoleptically acceptable throughout the seven day study and the pH remained relatively unchanged throughout the storage period, only varying slightly from the 6.5 initial pH to 6.4 at the end of seven days.

The initial water activity of the crumpet was 0.990 which, although not noted in the paper, is significantly higher than would normally be expected in crumpets manufactured in the UK (normally 0.96–0.98). The formulation also lacked any preservative agent which, in the UK, often includes vinegar (acetic acid).

Toxin was produced in all samples of high a_w, 'English-style' crumpet, irrespective of the storage conditions or whether pre- or post-bake inoculated (Table 4.35). In the post-bake inoculated product toxin was produced after 6 days, 5 days and 4.5 days in air, air + oxygen absorbent and CO_2/N_2 mixture, respectively. In the pre-bake inoculated product a similar pattern of toxin production was seen with toxin produced after four days in CO_2/N_2 and air + oxygen absorbent and six days in air. Counts of *C. botulinum* increased from 5×10^2/g at inoculation to log 4.9–5.9 at point of toxin production.

Table 4.35 Toxin production by *C. botulinum* (four type A and five type B) inoculated at 5×10^2/g into high water activity 'English-style' crumpets (a_w 0.990, pH 6.5) after storage at 25°C, adapted from Daifas *et al.* (1999)

Storage atmosphere	Post-bake inoculated	Pre-bake inoculated
	Time to toxin (days)	
Air	6	6
Air + oxygen absorbent	5	4
CO_2/N_2 (60 : 40)	4.5	4

Daifas *et al*. (1999) also reported the results from studies which demonstrated that high water activity pizza crusts (a_w 0.960) supported the growth and toxin production by *C. botulinum* when stored at ambient temperature for six weeks, irrespective of gaseous atmosphere. In contrast, high moisture bagels (a_w 0.944) failed to support growth and toxin production by *C. botulinum*.

Fresh-filled pasta products normally have water activities in the region of 0.93–0.95 and it is the water activity that provides the key safety factor in these products. These products are usually vacuum-packed or occasionally modified atmosphere packed. The pH of the filling can vary, with some of the tomato-based, filled products having pH values of 5.0 while others have a pH between 6 and 7. The products are given extensive shelf lives of several months and are usually destined to be fully cooked. However, it must be recognised that while most cooking guidelines aim to ensure the product receives a full cook, many consumers will cook the product until the outer pasta component is only partially cooked and may consume it even if the centre filling is not hot, i.e. it may not receive a full cook throughout the product.

Some research has been carried out into the growth of *C. botulinum* in filled and unfilled pasta products. Glass and Doyle (1991) inoculated commercial filled pasta products (meat or cheese filled tortellini) and flat noodle pasta products (linguine and fettucine) with proteolytic strains (five type A and five type B) of *C. botulinum* (1.1×10^3/g in flat noodles and 6.6×10^2/g in filled pasta). Laboratory prepared pasta products were also manufactured to various water activities and inoculated with *C. botulinum* (6–190/g in flat noodles and 150–510/g in filled pasta). Both commercial and laboratory prepared pasta products were then vacuum-packed and heated to 85°C for 15 minutes. The packs were then opened and placed under modified atmosphere conditions at 4°C or 30°C for 8–10 weeks. These packs were tested for toxin production at various intervals. In addition, to determine the heat stability of the toxin, some packs, which were shown to contain toxin after a predetermined time interval, were cooked according to the cooking instructions (placed in boiling water and boiled for one minute for flat noodles and eight minutes for filled pasta) and then re-tested for botulinum toxin.

Toxin was not detected in any sample of pasta (filled or unfilled) after storage at 4°C for 10 weeks (Table 4.36). Toxin was however detected in meat-filled tortellini at a_w values of 0.99 (pH not determined, 80% nitrogen, 20% carbon dioxide) and 0.95 (pH 6.2, 100% nitrogen) after one and six weeks respectively. In the same product, no toxin was detected at a_w

Table 4.36 Toxin production by proteolytic *C. botulinum* types A and B in filled and unfilled pasta, adapted from Glass and Doyle (1991)

Product	Time to toxin production (weeks)*	pH	Water activity	Gaseous atmosphere
Filled pasta				
Meat-filled tortellini	1	ND	0.99	80% N_2, 20% CO_2
Meat-filled tortellini	6	6.2	0.95	100% N_2
Meat-filled tortellini	Negative	6.2	0.93	100% N_2
Meat-filled tortellini	Negative	6.2	0.92	80% CO_2, 20% N_2
Cheese-filled tortellini	Negative	5.7	0.94	80% CO_2, 20% N_2
Unfilled pasta				
Linguine	2	6.9	0.96	80% N_2, 20% CO_2
Fettucine	Negative	6.6	0.95	100% N_2
Fettucine	Negative	6.6	0.93	100% N_2
Spinach fettucine	Negative	6.6	0.93	40% N_2, 60% CO_2

*Storage at 30°C for up to 10 weeks.
ND = not determined.

0.93 (pH 6.2, 100% nitrogen) or at 0.92 (pH 6.2, 80% carbon dioxide, 20% nitrogen) after 10 weeks' storage at 30°C. No toxin was detected in cheese-filled tortellini after 10 weeks at a_w 0.94 (pH 5.7, 80% carbon dioxide, 20% nitrogen).

In the unfilled products, toxin was detected after two weeks at 30°C in linguine (pH 6.9, 80% nitrogen, 20% carbon dioxide) at a_w of 0.96. No toxin however was detected in fettucine (pH 6.6, 100% nitrogen) at a_w 0.95 or 0.93, nor in spinach fettucine at a_w 0.93 (pH 6.6, 40% nitrogen, 60% carbon dioxide).

After cooking the toxic flat noodles (initial toxin concentration between 10^3 and 10^4 MLD/g) and the filled pasta (initial toxin concentration between 10^2 and 10^3 MLD/g), no toxin was detected.

In a similar experiment conducted by del Torre *et al.* (1998), a range of filled pasta products was inoculated with spores of proteolytic *C. botulinum* types A, B and F (mixture of eight strains) at levels of 5×10^2–1×10^3/ pasta piece and stored under 83% nitrogen, 15% carbon dioxide and 2% oxygen for 50 days at 12°C and 20°C. No toxin was detected in any product stored at 12°C. Artichoke-filled pasta (pH 5.6, a_w 0.96) stored at 20°C did not support the development of toxin. However, meat-filled pasta (pH 6.2, a_w 0.95), ricotta/spinach pasta (pH 5.9, a_w 0.95) and salmon pasta (pH 6.1, a_w 0.95) all allowed the development of botulinum toxin after 50 days, 50 days and 30 days respectively (Table 4.37).

The crucial role of water activity in the safety of fresh pasta products is clear and, either alone or in combination with pH, should be considered carefully in any hazard analysis of product formulation and process conditions required for consistent achievement of safety with respect to *C. botulinum*.

A number of these types of products contain preservative agents such as potassium sorbate or other organic acids. Although some research has shown inhibition of *C. botulinum* with different preservatives, contradictory findings have also been reported where the preservative has little or no effect. In addition, organic acids are far more effective at low pH values, usually < 5, where they exist predominantly in the undissociated form. As these values are rarely approached in bread and pasta products, it is unlikely that the preservatives would have a significant effect, particularly against *C. botulinum*. It is recommended that if these types of preservatives are to be relied upon to control the growth of *C. botulinum* in a pasta or bread product, then a challenge test should be designed and

Table 4.37 Toxin production by proteolytic *C. botulinum* types A, B and F in filled pasta, adapted from del Torre *et al.* (1998)

Product*	Toxin production (number of products toxic/number examined)	Time to toxin production at 20°C (days)	pH	Water activity
Artichoke-filled tortelli	Negative (0/5)	–	5.6	0.96
Meat-filled tortelli	Positive (1/5)	50 (negative at 30)	6.2	0.95
Ricotta and spinach-filled tortelli	Positive (3/5)	50 (negative at 30)	5.9	0.95
Salmon tortelli	Positive (2/5)	30 (negative at 15)	6.1	0.95

*Products inoculated with 5×10^2 spores/pasta parcel (meat filled) and 1×10^3 spores/pasta parcel (other fillings) and stored at 20°C under 83% N_2, 15% CO_2 and 2% O_2 for up to 50 days.
Toxin was not detected in any pasta stored at 12°C.

carried out to confirm that the preservative is providing a protective effect under the conditions present in the final product formulation.

It is clear from the work reported by a number of researchers that flour-based products such as breads and pasta do support the growth of and toxin production by *C. botulinum* providing the conditions in the product are suitable. Products of this nature however have not yet been linked to outbreaks of botulism and this historical precedence has been used by some as a means to justify the long shelf lives under ambient conditions given to these products.

However, a good history of safety does not protect the product from the future potential to cause outbreaks, if appropriate conditions arise. The development of higher moisture, anaerobically packed products of this nature, stored for long periods at ambient temperatures is a relatively new advance and suitable controls must be put in place to prevent outbreaks occurring in this vulnerable product sector.

It is recommended that if products of this nature are to be allocated extended shelf lives they should be formulated to water activity values of 0.94 or less throughout the product. Any product with water activities of 0.95 or greater in any component should have its shelf life restricted or should have an appropriate challenge test carried out on the proposed finished product to establish a safe shelf life with regard to *C. botulinum*. The safety of products of this nature should be readily achievable using sound HACCP based approaches; this will ensure that high quality products continue to be developed which are designed to be safe and stable.

SALAMIS AND RAW, DRY-CURED MEATS

Cured, fermented meat products, e.g. salamis, have a good safety record with respect to botulism with no confirmed outbreaks. Outbreaks have been reported associated with raw, dry-cured meat, usually involving small-scale or home manufacture. Salamis and raw, dry-cured meat products are mainly produced using traditional processes and, if inadequately controlled, have significant potential to cause botulism outbreaks due to their manufacturing process. Salamis include products manufactured in many countries including Germany, Denmark and Italy. The premium end of the market however is commanded by the raw, dry-cured meats, with products like Parma ham having a world-wide distribution. These products are made using long established traditional processes and the countries most commonly associated with their manufacture are those in continental Europe including Italy, France and Germany. However, the manufacturing processes for salamis and raw, dry-cured meats differ quite markedly.

Description of process

Salamis are made by bowl chopping raw meats and mixing them with a variety of ingredients including fat, herbs, spices, salt, fermentable sugar and the preservative sodium nitrite (Figure 4.9). Most salamis have starter cultures added to the raw meat mix to facilitate the subsequent fermentation stage, but some traditional processes rely on the development of natural lactic microflora for their fermentation. Pork is the predominant meat used in the manufacture of salamis but it is also often mixed with beef and other meat species are beginning to be used, e.g. poultry meat.

The process involves taking the meat mixture from the bowl chopper and then forcing this into casings to form a sausage shape. Casings may be of natural animal origin or made from artificial edible materials. The raw salami sausage is then sealed at both ends of the sausage using a metal crimp or with string. Each sausage is then usually suspended on a metal rack with an entire batch of salami prior to being placed in the fermentation chamber. The batch is then subjected to controlled conditions of temperature and humidity which allow an active bacterial fermentation to occur alongside the commencement of a drying process. The conditions for fermentation and drying vary between different countries and between different processors within the same country.

In European salami manufacture the fermentation stages tend to be at lower temperatures (20–30°C) for longer periods of time, whereas in the

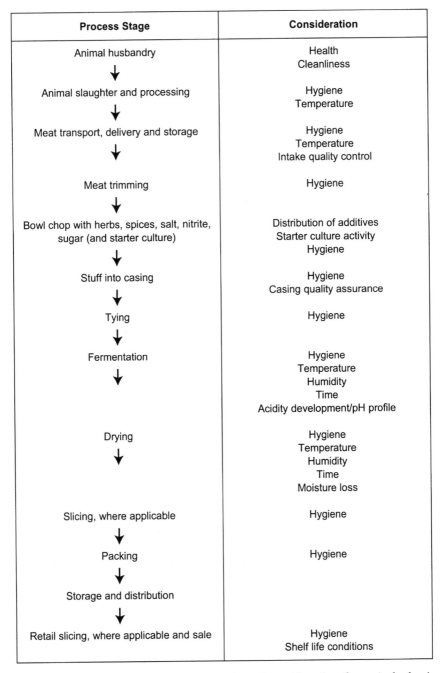

Process Stage	Consideration
Animal husbandry	Health Cleanliness
Animal slaughter and processing	Hygiene Temperature
Meat transport, delivery and storage	Hygiene Temperature Intake quality control
Meat trimming	Hygiene
Bowl chop with herbs, spices, salt, nitrite, sugar (and starter culture)	Distribution of additives Starter culture activity Hygiene
Stuff into casing	Hygiene Casing quality assurance
Tying	Hygiene
Fermentation	Hygiene Temperature Humidity Time Acidity development/pH profile
Drying	Hygiene Temperature Humidity Time Moisture loss
Slicing, where applicable	Hygiene
Packing	Hygiene
Storage and distribution	
Retail slicing, where applicable and sale	Hygiene Shelf life conditions

Figure 4.9 Process flow diagram and technical considerations for typical salami.

USA, processes are operated at higher temperatures ($> 30°C$) for shorter periods. The active growth and metabolic activity of the starter cultures results in the reduction of pH and increase in organic acid content of the meat mix. The pH levels decrease to 5.0 or below and the safety of the process in relation to organisms like *C. botulinum* undoubtedly depends on factors including rapid acidification together with the presence of other inhibitory agents and reduction in moisture levels.

Drying the salami occurs both during the fermentation stage, which can last for 3–5 days, and also during the subsequent ripening process where the product is maintained at temperatures of approximately 15°C or less for several weeks or months during which moisture is lost and water activity is reduced. The water activity of the salami reduces from approximately 0.98 at the beginning of the process to below 0.94 at the end, although many salamis may have water activities significantly lower than 0.90.

A number of European salami processes include a mould-ripening stage which allows the development of mould growth on the outer surface of the sausage during the drying stage. This imparts a distinctive flavour to the product but also accounts for a significant increase in pH, from the 5.0 or less that it achieves during the fermentation stage to pH 6–7 after the mould-ripening stage. Of course, this can vary markedly across the salami depending on its diameter, with the lowest pH being in the centre, furthest away from mould growth, and the highest being at the surface. The mould may be left on the product prior to packing or it may be washed or brushed off for certain markets.

Some salamis may be smoked prior to the drying stage by holding the product for a time in smoking rooms filled with natural wood smoke.

Products may be sold without wrapping but many these days are film-packed and either gas flushed or vacuum-packed. Products packed as whole salamis are usually destined for sale on delicatessen counters where they will be sliced on the counter, but they may also be sliced at the production site and supplied for sale as pre-packs.

Salami products often have extremely long shelf lives because the harsh conditions usually present in the finished salamis (low pH, low water activity and the presence of antimicrobial preservatives) preclude growth of most microorganisms. Final product salamis usually have a pH between 5 and 7 depending on whether the product is mould-ripened and at what position in the salami the pH is measured. The water activity can be as low as 0.85 but is usually closer to 0.90, with an aqueous salt content of 5–10%. Although most salamis are in themselves generally safe to display and sell under ambient conditions they are often sold under refrigeration conditions.

Raw, dry-cured meats are manufactured by quite different procedures to those used for salamis (Figure 4.10). The meat, usually pork but sometimes beef, is used as whole anatomical pieces of meat and undergoes only minor trimming to achieve the correct fat content. The meats are not comminuted. The meat pieces are usually dry-salted by rubbing salt,

Process Stage	Consideration
Animal husbandry ↓	Health Cleanliness
Animal slaughter and processing ↓	Hygiene Temperature
Meat transport, delivery and storage ↓	Hygiene Temperature Intake quality control
Meat trimming ↓	Hygiene
Salting and storage ↓	Distribution of salt Hygiene Temperature
Drying ↓	Hygiene Time Temperature Humidity Moisture loss
Slicing, where applicable ↓	Hygiene
Packing ↓	Hygiene
Distribution ↓	
Retail slicing, where applicable and sale	Hygiene

Figure 4.10   Process flow diagram and technical considerations for typical raw, dry-cured meat.

which sometimes includes sodium nitrite, over the surfaces and then placed in chilled maturation rooms at <5°C to encourage the development of the natural lactic microflora on the meat surfaces. During several days or weeks of storage, the meat is regularly removed, re-salted and turned to ensure even distribution of salt throughout the raw meat. Following the refrigerated storage, the salted raw meats are dried at temperatures of <15°C for several months. This stage may be accompanied by a mild fermentation as a result of growth of lactic microorganisms, although the pH only decreases slightly. Some products may be smoked prior to drying. After drying the product usually achieves a pH between 5 and 6 and a water activity of 0.85–0.90. Like salamis, raw, dry-cured meat products may be sold as whole pieces or sliced for pre-pack sale.

The hazard presented to salamis and raw, dry-cured meats by *C. botulinum* is exacerbated in salamis as the contaminants on the surface of the meat become distributed throughout the product during bowl chopping and therefore, to achieve and maintain safety, antimicrobial factors must be present in adequate concentrations in all parts of the product. In addition, the temperatures employed in the manufacture of salamis are higher than those used for raw, dry-cured meats and lend themselves to more rapid growth of the organism. However, in contrast, raw, dry-cured meats are sometimes made without the anti-botulinum agent, sodium nitrite, and, with less acidity development, they may be susceptible to the growth of the organism if suitable conditions prevail. In addition, the size of the meat pieces can be large, e.g. whole legs of pork, and the slow penetration of salts can present an additional opportunity for growth.

Although most emphasis in the following text will be placed on salamis, the concern relating to the potential for growth of *C. botulinum* in raw, dry-cured meats will also be considered.

Raw material issues and control

The major raw ingredient in the manufacture of salamis and raw, dry-cured meats is the raw meat itself. Several species of meat are used in these products with the most common being pork. *C. botulinum* is known to be a contaminant of raw meat and most surveys tend to report the highest incidence in pork. While good conditions of hygiene during slaughter, evisceration and primal processing clearly impact on reducing the extent of contamination, it is inevitable that the raw meat material will occasionally be contaminated with *C. botulinum*. The organism may be present in soil or faeces which contaminates the skin/hide of the animal.

These are then transferred to the carcass during slaughter by cross-contamination from personnel handling practices, the environment and equipment surfaces and utensils. It is important to ensure that complacent practices do not allow the extensive build-up of microbial loads in the environment and consequent cross-contamination, but it must be recognised that it is not possible to completely exclude the presence of *C. botulinum* from the raw incoming meat.

Stocks of raw meat used in the production of these products are usually held frozen or chilled and control of chill storage times and temperatures of the raw meat is important to preclude opportunities for extensive growth of non-proteolytic strains of *C. botulinum*. In most cases the quality of the finished product requires the raw meat to be as fresh as possible and, if chilled, to be used within 48–72 hours from receipt. Care must be taken when using frozen meat to ensure that tempering of the meat to chill temperatures is conducted without exposing the material to abuse temperatures. To speed up defrosting, some producers may leave the frozen meat at ambient temperatures, but this is clearly not an acceptable practice as temperatures of the meat are uncontrolled and will potentially allow the growth of contaminating microorganisms. Frozen meat should be defrosted under controlled, chilled temperature conditions to avoid temperature abuse.

While raw meat is the primary raw ingredient for these products it should not be forgotten that a variety of other ingredients are also added to salamis. These will include salt and sodium nitrite and may also include pepper and a variety of other herbs and spices, most of which are added in their dry form. Herbs and spices are known to be common sources of spore-bearing bacteria which arise as contaminants from the soil, dust and air in which the products are grown and harvested. It may be of some benefit to monitor the raw material herbs and spices for microbial contamination including spore loads to ensure that poor quality batches are not being received. While it is difficult to ensure that such materials are always of the highest quality, it may at least be possible to ensure that the poorest quality materials are not used in these types of products.

Raw, dry-cured meats usually only have salt and the curing agents, sodium nitrite and/or sodium nitrate, as ingredients and these are usually restricted to the external surfaces only and in dry form. Such material represents little or no additional hazard in relation to contributing to the potential presence of *C. botulinum*.

Process issues and control

Salamis

Salamis are manufactured by bowl chopping the raw meat and mixing in various ingredients to form the salami paste (although termed a paste, this can include significantly different meat piece sizes from very large, i.e. coarse, to very small, i.e. fine). This is then forced into casings to form the sausage shaped salami stick. The size of the salami is dictated by the diameter of the casing and the points at which the salami is tied or sealed using string or metal crimps.

The bowl chopping stage presents a significant opportunity for distributing microbial contamination throughout the meat mix. Any contamination on the external surfaces of the raw meat is distributed throughout the meat mix during bowl chopping and this is then further mixed with contaminants introduced from the herb and spice additions. The widespread distribution of microorganisms within the mix is generally considered to create the greatest hazard with respect to these types of products as it increases the potential for the organisms to be distributed to points in the product where antimicrobial factors may be less effective than if the bacteria were restricted to the product surfaces.

It is important to prevent poor cleaning and hygiene practices from allowing build-up of contamination in and on the meat mix processing equipment and particularly the bowl chopping and casing/filling equipment. Some areas within such equipment are notoriously difficult to clean effectively and it is important to ensure that poor cleaning practices do not compromise successive batches of product by allowing contamination to build up and be passed on. Although such contamination is more likely to consist of organisms other than *C. botulinum*, this should not detract from efforts to clean equipment and utensils effectively.

The most critical aspect of this initial stage of the salami process is to ensure that all factors likely to contribute to the subsequent control of bacterial pathogens, including *C. botulinum*, are distributed homogeneously throughout the meat mix. It is critical that salt, nitrite and nitrate, if added, are distributed evenly and at the correct ingoing concentration. Failure to do this may create local areas in the salami where levels of inhibitory factors are inadequate. If starter cultures are added to the meat mix, it is critical that such cultures are added at the right level and mixed in thoroughly. It is also important to ensure that the culture is active and if fermentable sugars are required that they are present at the correct level. Addition of ingredients necessary for controlling product safety must be recorded and it is usual to monitor the distribution of the controlling factors such as salt and nitrite by chemical checks of the salami stick after filling and during the subsequent process.

One of the most critical stages in the manufacture of salami is the fermentation stage. Salamis are usually freely suspended from a metal frame/rack and then placed in a chamber. Hanging and spacing on the frame allows free circulation of air and effective moisture loss.

The chamber is usually controlled to precise levels of humidity and temperature and the process times and temperature are designed to encourage the development of the lactic microflora in the meat mix.

Salami processes can differ markedly, but in Europe they tend to be at temperatures of 20-25°C for a duration of 3-5 days. In the USA, the fermentation tends to be at higher temperatures ($>30°C$) for shorter duration (1-2 days). In both cases the objective of fermentation is to achieve growth of lactic microflora to produce inhibitory organic acids and reduce pH while at the same time, through humidity control, reducing the moisture content of the meat.

With respect to *C. botulinum*, the presence of salt and nitrite in the meat mix will undoubtedly help to restrict the growth of the organism during fermentation. However, the role of nitrite has been questioned by a number of workers who have not seen any beneficial effect regarding inhibition of *C. botulinum* by its presence versus absence in challenge test studies (Lücke and Roberts, 1993).

Without an active fermentation and rapid drying within the meat mix it is clearly possible for the organism to grow very quickly under the process conditions used. The initial meat mix will have a water activity of c. 0.97–0.98, aqueous salt concentration of 3%-5% and a pH varying from 5.8 to 6.5 depending on the meat species and source. Ingoing nitrite is usually 75-150 ppm. These inhibitory factors would not individually be expected to completely prevent the growth of proteolytic strains of *C. botulinum* and it is likely that the fermentation by-products of the lactic acid bacterial growth together with the reduction in pH are important factors in making the fermentation stage of the salami process safe.

During fermentation, the pH of the salami can decrease to < 5.0, although some varieties may not achieve this pH while others may decrease to as low as pH 4.5. This will be significantly affected by the presence of fermentable sugar and the addition of starter cultures as well as the times and temperature employed in the fermentation stage of the process. Clearly, the use of added starter cultures can make the fermentation process more controllable but, irrespective of this, the critical aspect of this stage is the rate of production of organic acids, usually monitored by acidity development or pH reduction. As the product loses 10-15% moisture during this stage, this will also be important in relation to safety as the added inhibitory factors such as salt and nitrite, together with the organic acids produced by the growth of the lactic microflora, all become more concentrated in a reducing aqueous phase. Therefore, control of the required changes in all of the factors necessary for safety is critical and any process modification, e.g. to reduce the fermentation activity or level of inhibitory factors, needs careful consideration in relation to its potential effect on growth of contaminating pathogenic bacteria.

Several studies have been carried out on the potential for growth and toxin formation by *C. botulinum* during the manufacture of salamis. In general, it is clear that providing active fermentation takes place through the growth of natural lactic microflora or added starter bacteria, with the concomitant production of organic acids and pH decrease, then *C. botulinum* will not grow to toxin-forming levels. It is, however, also apparent that the exact physico-chemical boundaries that could allow growth and toxin formation are not well established and therefore the degree of safety afforded by such processes can be a matter of subjective assessment. Clearly, products achieving the following characteristics are likely to result in less risk of allowing germination and growth of *C. botulinum*:

- starter cultures used
- rapid fermentation (e.g. pH < 5.0 within 24 hours)
- high initial salt content (3%-5% aqueous phase)/low initial water activity (< 0.97)
- presence of nitrite
- raw materials with low bacterial spore contamination levels.

Typical changes occurring during a salami manufacturing process are detailed in Table 4.38.

Salami products are usually considered to be of greatest concern with respect to the growth and toxin production by *Staphylococcus aureus* and criteria have been established by the American Meat Institute to determine safe fermentation conditions to control this organism based on the temperature and time taken to reach a certain pH value using added

Table 4.38 Changes in controlling factors during typical European salami manufacturing process

Controlling factors	Process stage					
	Raw meat paste	Fermentation			Drying	
Time (days)	0	1-3	4-5	6-10	11-30	31->50
Temperature (°C)	20-25	15-20	15-20	10-12	8-12	8-12
a_w	0.96-0.98	0.95-0.96	0.94-0.96	0.93-0.94	0.90-0.93	0.88-0.92
pH	5.7-6.0	5.2-5.5	5.0-5.3	4.9-5.3	4.9-5.5*	4.9->6.0*
Weight loss (%)†	0	5-10	10-15	15-20	25-30	30-35
Aqueous salt (%)	3-5	—	—	—	—	7-9

*pH may increase due to mould ripening of some varieties.
†Weight loss is a measure of the moisture reduction with initial moisture levels ranging from 55 to 65%.

starter cultures (American Meat Institute, 1989). The process is defined in terms of degree-hours which represents the number of hours of the fermentation process multiplied by the number of degrees (°F) above 60°F. The maximum number of degree-hours to generate a safe product with respect to *S. aureus* has been calculated for a variety of fixed-temperature fermentations (Table 4.39). If a process takes longer than the specified degree-hours to achieve a pH of 5.3 or less then the potential for *S. aureus* growth and toxin production is much greater. Provided this level is achieved by the end of the fermentation stage, it is probable that not only will the growth of *S. aureus* be controlled but also that of other pathogens (Incze, 1992) including *C. botulinum*. However, it is also considered important that a low water activity is also achieved within this fermentation period (i.e. < 0.95) to provide an additional safeguard against the growth of *C. botulinum* during successive stages.

Table 4.39   Recommended fermentation conditions to control fermented meat manufacture, adapted from American Meat Institute (1989)

Temperature of fermentation	Degree-hours* (maximum) to reach pH 5.3	Maximum permissible fermentation time at each temperature (hours)
75°F	1200	80
80°F	1200	60
85°F	1200	48
90°F	1000	33
95°F	1000	28
100°F	1000	25
105°F	900	20
110°F	900	18

*Degree-hours are calculated by multiplying the fermentation time (time to reach pH 5.3 or less) by the number of degrees above 60°F employed for the fermentation.

Challenge tests conducted on properly fermented and dried salamis have failed to demonstrate toxin production by *C. botulinum*. However, where the fermentation was artificially depressed and where elevated temperatures were maintained, toxin production was demonstrated. Interestingly this occurred both in the presence and absence of nitrate or nitrite.

The importance of pH reduction and the presence of dextrose, to facilitate this, has been shown by Christiansen *et al.* (1975). They inoculated a summer sausage batter mix with spores of *C. botulinum*. In one series of experiments they examined the ability of strains of *C. botulinum* (100 and 10 000 spores/g meat mix) to grow and produce toxin when fermentation occurred normally. A meat mixture including 2.5% salt, 2% dextrose and 0.35% spices was supplemented with varying concentrations of nitrite (0, 75, 150 and 300 ppm) and nitrate (0 and 1500 ppm) and fermented both with and without a starter culture. Fermentation conditions for the starter culture inoculated product included the following; two days at 10°C followed by 4 h at 32.2°C, 4 h at 32.2°C only, or 32.2°C until a pH of 5.3–5.4 was reached (18–24 h). All products were then heated to 58.3°C, sliced, vacuum-packed and then stored at 27°C for up to 28 days. Product formulated without a starter culture was fermented at 32.2°C for 18 h and then heated to 58.3°C prior to slicing, packing and storing at 27°C. All products had a pH of 5.2–5.8 after the fermentation stage which decreased to 4.6–4.9 after 8 days' storage at 27°C. Toxin was not produced in any product after the fermentation stage and storage at 27°C, even though a drying stage, an essential component of salami production processes, was omitted. Control of *C. botulinum* was attributed to an effective fermentation which occurred in all products to cause pH reduction by the growth of either the added starter culture or, in the absence of added starter, the natural lactic microflora in the product. Inhibition was clearly not reliant on nitrite, nitrate or added starter culture but did rely on the presence of salt and particularly dextrose in the formulation, the latter acting as the substrate to facilitate pH reduction.

In a second experiment, Christiansen *et al.* (1975) omitted the fermentation step and examined the effect of the presence or absence of dextrose and starter culture with varying concentrations of nitrite (0, 50, 100 and 150 ppm) on toxin production by strains of *C. botulinum* (mixture of five type A and five type B). The product formulation was the same as that already described and after processing it was heated to 58.3°C, sliced, packed and stored at 27°C.

The mild heat process had little effect on the levels of starter bacteria in the product as demonstrated by plate counts before and after heating. These were able, during subsequent product storage, to utilise suitable substrates present and produce organic acids, thus lowering pH.

Products with dextrose added showed marked pH reduction during storage at 27°C (Table 4.40). In contrast, pH remained high in products formulated without dextrose. There was little effect noted of nitrite or starter culture addition on the pH achieved.

The production of toxin was completely inhibited in all formulations where dextrose was present and where nitrite was at levels of 50 ppm or more, even in the absence of a controlled fermentation stage (Table 4.41).

Table 4.40 Effect of dextrose on pH change in product where the initial fermentation stage was omitted, adapted from Christiansen et al. (1975)

pH change (average)		Before processing	After processing	1 week at 27°C	2 weeks at 27°C
	With dextrose	5.63	5.97	4.68	4.38
	Without dextrose	5.58	5.87	5.74	5.86

Process: ingredients mixed, product formed, heat process applied (to reach 58.3°C), cooled, sliced, packed and stored.

Table 4.41 Growth and toxin production by *C. botulinum* in summer sausage where the initial fermentation stage has been omitted and under varying concentrations of nitrite, in the presence or absence of dextrose and added starter cultures, adapted from Christiansen *et al.* (1975)

	Nitrite (ppm)	Starter culture*	Dextrose*	Earliest time to toxin production (days)
Formulations toxic after storage at 27°C for up to 112 days	0	Y	Y	56
	0	Y	N	14
	0	N	Y	7
	0	N	N	14
	50	Y	N	14
	50	N	N	14
	100	Y	N	28
	150	Y	N	21
	150	N	N	21
Formulations not toxic after storage at 27°C for up to 112 days	50–150	Y/N	Y	

*Present: Y = Yes; N = No.

This was attributed to the combined effect of decreased pH afforded by the presence of dextrose as a substrate for lactic acid bacterial activity and also the inhibitory properties of nitrite. Starter culture and dextrose alone did not prevent toxin formation, nor did nitrite in isolation at concentrations up to 150 ppm. However, fewer samples became toxic when starter cultures were added but their presence did not prevent toxin formation completely. Clearly the use of dextrose, nitrite and starter culture together appear to offer the greatest protection against the potential development of *C. botulinum* in fermented meat products.

Similar findings were reported by Kueper and Trelease (1974) who found no toxin production by *C. botulinum* (types A and B) in meat undergoing an effective fermentation with or without nitrite, nitrate or starter culture and when the products were heated to 137°F (58.3°C) after fermentation and then packed and stored at 45°F (7.2°C) and 80°F (26.7°C) for up to three months. The products were formulated to 53.2% moisture, 2.6% salt with 24.8% fat and 2% dextrose and achieved a pH of 5.6 and 5.2 after initial fermentation. Without a fermentation stage, they found the combination of dextrose (2%) and nitrite (50 ppm) to be essential for controlling *C. botulinum* (Table 4.42).

Nordal and Gudding (1975) inoculated a salami mix with *C. botulinum* types B and E at levels of $> 10^5$/g and concluded that no toxin production would occur if the initial water activity was 0.97 and the pH was 5.7, with further security being obtained by the use of a starter culture and components facilitating pH reduction such as a fermentable sugar and glucono-delta-lactone.

Lücke *et al.* (1983) investigated the fate of *C. botulinum* (proteolytic and non-proteolytic strains) inoculated at 10^4 spores/g of mix in two typical fermented meat products (salami and teewurst), with and without nitrite addition. Salami was formulated to 40% pork, 30% beef, 30% pork back fat to which was added 2.8% curing salt containing variable concentrations of nitrite and ascorbate, 0.3% glucose, 0.3% sucrose and 0.35% spice. Teewurst was formulated with the same meat mix but with the addition of 2.4% curing salt, 0.3% mix of starch hydrosylate and 0.32% spice.

Salamis were formulated to contain no nitrite, 40 mg/kg nitrite + 550 mg/kg sodium ascorbate or 115 mg/kg nitrite only. Products were fermented/ripened at 15°C in one experiment and at 25°C for seven days followed by 20°C for periods up to five weeks in a second experiment. No starter cultures were added. The initial pH of the salami was 5.6–5.8 and the initial water activity 0.96.

Table 4.42 Toxin production by *C. botulinum* types A and B in a fermented sausage where the initial fermentation stage was omitted, adapted from Kueper and Trelease (1974)

Nitrite (mg/kg)	Toxin production during storage at 80°F for up to 16 weeks			
	With culture/ with dextrose	Without culture/ with dextrose	With culture/ without dextrose	Without culture/ without dextrose
0	Toxin (pH 4.4)	Toxin (pH 4.4)	Toxin (pH 6.1)	Toxin (pH 5.8)
50	No toxin (pH 4.32)	No toxin (pH 4.4)	Toxin (pH 6.05)	Toxin (pH 5.85)
150	No toxin (pH 4.32)	No toxin (pH 4.5)	Toxin (pH 5.85)	Toxin (pH 5.6)

( ) pH after two weeks at 80°F (26.7°C).

Teewurst was prepared with no nitrite, with 50 mg/kg nitrite + 550 mg/kg ascorbate, with 94 mg/kg nitrite or with 115 mg/kg nitrite. The product was fermented at 25°C for seven days and then ripened at 20°C for up to five weeks. Conditions in the chamber were maintained to ensure the product pH remained above 5.2 and the water activity above 0.94 during the fermentation. The initial water activity was 0.965 and the pH was 5.8.

During salami manufacture the lactobacilli, initially present at low levels (c. 100/g), rapidly became the dominant microflora and increased to in excess of 10^8/g within eight days at 15°C and three days at 25°C. The pH decreased to c. 5.0 during this period and the water activity decreased to 0.94–0.95. Little difference was recorded in the pH of any of the formulations and no toxin was detected, even in the absence of nitrite. Also noted was the low possibility of growth and toxin production by *C. botulinum* in salami given a reasonably low initial water activity (0.965) and low pH (5.8 or less). In the teewurst product, it was found that no toxin was produced in any formulation.

In both salami and teewurst, the researchers noted that formulations containing no nitrite did not produce an organoleptically acceptable product, with the production of rancid flavours due to auto-oxidative processes.

Raw, dry-cured meats

Raw, dry-cured meats differ significantly from salamis in their manufacture which makes the inherent hazard associated with the growth of *C. botulinum* somewhat different from that in the traditional salami process. Raw, dry-cured meats are usually manufactured from whole anatomical pieces of meat with contaminants usually located and restricted to the external surfaces of the meat. Although contaminants can enter the deep muscle tissues, particularly through poor hygiene standards during slaughter using contaminated knives to cut the animal's throat leading to transfer of bacteria through the blood supply, this is likely to occur only rarely.

Raw, dry-cured meats such as Parma ham are manufactured from whole legs of pork. For these products, the skin remains on the flesh and contamination will primarily be restricted to the rind and the exposed meat surface.

Following meat preparation, e.g. trimming, the first stage in the process of manufacture of these products is to apply salt, usually by hand to the external meat surfaces. The salt, including nitrite and/or nitrate, is rubbed onto the surface of the meat, leaving large amounts of salt crystals on the surface. Some dry-cured meats also have other flavourings added to the curing mixture applied to the meat.

After salting, the meat is either hung up or stacked on top of each other in holding containers and left at low temperatures (usually <5°C). Salting is repeated at various intervals over a period of several weeks during which time the meat is also turned regularly. The high salt content on the meat and the low temperature tends to select a dominant microflora of lactic acid bacteria.

During this stage the salt diffuses into the centre of the meat raising the aqueous salt content to approximately 5% or more. The product is then usually suspended on racks and subjected to a drying process initially at temperatures of 10–15°C for several months. Some products may be dried at higher temperatures, up to 20–25°C, although this is usually after a period of drying at a lower temperature.

During the drying stage, the product undergoes a mild fermentation caused by the growth of lactic acid bacteria, although this usually results in little reduction in pH and production of organic acids. The product also loses a significant amount of moisture during this stage and the water activity falls from approximately 0.95–0.96 in the primary stages of drying to 0.92 or below in the latter stages of drying. At the end of the manufacturing process most of these types of product have aqueous salt values above 7%–8% and water activities below 0.90. Typical changes in raw, dry-cured meat manufacture are shown in Table 4.43.

Table 4.43 Changes in controlling factors during typical raw, dry-cured meat manufacture

Controlling factors	Stage			
	Raw meat	Salting	Maturation	Drying
Time	Day 0	Week 3–4	Week 4–10	Week 10–30
Temperature (°C)	1–4	1–4	1–4	10–20
a_w	0.97–0.98	0.95–0.96	0.94–0.95	0.87–0.92
pH	5.7–6.0	5.7–6.0	5.7–6.0	5.5–5.8
Weight loss (%)*	0	1–2	15–20	30–35
Aqueous salt (%)	0	4–5	7–8	9–11

*Weight loss is a measure of the moisture reduction with initial moisture levels ranging from 60% to 70%

It is often difficult to clearly rationalise the exact factors in the process of raw, dry-cured meat manufacture that afford safety with regard to *C. botulinum*, especially in those products where salt is the only added preservative factor. Clearly, the organism will be present as a contaminant on occasion and therefore growth must be controlled.

It is probable that, as the organism is most likely to be present on the external surfaces of the meat, this coincides with the harshest conditions of preservative factors. Levels of salt at the surfaces are likely to exceed those capable of allowing growth of the organism, i.e. 10% or greater aqueous salt. During the early stages of the process, growth of *C. botulinum* would be restricted to non-proteolytic types due to the chilled storage conditions used. Maintaining low temperatures is therefore vital while salt equilibration occurs in the meat and while the development of lactic microflora is encouraged.

Lücke and Roberts (1993) reported on studies of bone-in hams inoculated with spores of proteolytic and non-proteolytic strains of *C. botulinum*. Proteolytic strains did not produce toxin in a typical curing process but toxin was produced by non-proteolytic *C. botulinum* types B and E when salting was carried out at 8°C but not at 5°C. Toxin was produced during the drying stage if the internal water activity was above 0.97 and no protective effect was observed in relation to the use of nitrite or nitrate.

Upon elevation of temperature, the potential exists for growth of proteolytic and non-proteolytic strains. Control of non-proteolytic strains, even under ambient temperature conditions, is likely to be achieved by the levels of salt in the product, which will mostly exceed 5% (aqueous) by the time of temperature elevation. However, they will rarely achieve 10% or greater throughout the meat muscle and therefore proteolytic strains may be capable of growth. pH levels may also be decreasing slightly, although this is likely to be more marked on the outside of the product.

It is clear that the combination of water activity and pH during the drying stage may restrict the growth of proteolytic strains of *C. botulinum*. For example, Baird-Parker and Freame (1967) found growth of vegetative cells of both proteolytic and non-proteolytic *C. botulinum* types A, B and E could be prevented at 20°C when the pH was 5.5 and the a_w 0.97 or less, although at 30°C growth was only prevented when the a_w was < 0.97 at pH 5.5. However, also at 30°C, when the pH was 6.0, the type B strain tested could only be inhibited at an a_w of < 0.96. Therefore, it is likely that the combination of low pH meat (< 6.0) together with decreased a_w (< 0.96) achieved during salting play an important role in restricting growth of *C.*

botulinum during the subsequent drying stages. Indeed, Lücke and Roberts (1993) referred to recommendations that raw meat for curing should have a pH below 5.8 and salting should both be conducted at 5°C or less and achieve an internal a_w below 0.96 before elevation of temperature.

It is also important to recognise that the first stage of most drying phases for these types of products involves increasing temperature to low ambient levels (10–15°C) for a period of time and it is likely that the moisture loss during this stage and consequent depression in a_w and increase in aqueous salt occurs faster than the growth of any contaminating strains of *C. botulinum*. Upon subsequent elevation to higher temperatures, the water activity and low pH may be sufficient to prevent growth of proteolytic and non-proteolytic strains and ensure safety.

As drying of salamis and raw, dry-cured meats results in loss of moisture from the outer regions of the meat first, it is possible that conditions of rapid drying or insufficient drying may lead to higher moisture levels in the centre of the product. Therefore, ensuring that drying schedules are strictly adhered to is critical to the safety of the product during processing and also during the ensuing shelf life.

Many salamis and raw, dry-cured meats are highly traditional products, and the processes of manufacture would be difficult to modify for associated reasons. The key to the safety of these products is an understanding by the manufacturer of the factors inherent in the process and product that affect microbial survival and growth. Challenge test studies, to determine the changes occurring in controlling factors such as pH reduction, acidity development, moisture loss and water activity decrease throughout the process and the effect these changes have on the microbial population including *C. botulinum*, are strongly recommended. Such challenge studies should be undertaken by experienced microbiologists using experimental facilities. Understanding these changes and effects will allow the key process changes that are shown to control the organism to be incorporated as routine process controls so that any process which does not subsequently achieve the required changes in the allotted time can be highlighted for remedial attention, as the chances of pathogen growth may be greater in these batches.

Final product issues and control

Salamis and raw, dry-cured meats are usually supplied in bulk form, often after vacuum packaging, or they may be sliced and sold as pre-packs. The final products have a very low water activity, usually < 0.90, high aqueous salt levels (5–10%) and low pH (< 5.8), although all of these factors vary considerably depending on the individual product type.

The vast majority of these types of product would be considered to be essentially microbiologically stable in the final product form and certainly growth of *C. botulinum* under chilled conditions would not be expected. Indeed, even storage at ambient temperatures would not allow growth of the organism in most products based on conditions of water activity alone, irrespective of other controlling factors such as pH, acidity and preservatives, e.g. nitrite. Clearly, it is important if retailing such products under ambient conditions to ensure that the controlling factors are appropriate and are consistent throughout the product.

Shelf life allocated to finished product is usually quite long and is more dictated by the organoleptic quality than safety. Pre-pack products which may be whole or sliced are sold in vacuum or modified atmosphere packages under refrigeration with lives of several weeks or more. Alternatively, whole dried meats or salamis may be sliced on the delicatessen counter for sale to the customer on demand and, because of excessive drying and loss of quality, such sliced product is usually allocated a shelf life of less than a week.

Some salamis may have post-process additives such as herbs or spices applied to the surface. This is usually achieved by coating the salami in gelatin and then dipping into peppercorns or other material to achieve a coating. Other salamis may be waxed and it is clearly important that the product has been fully dried prior to this process as this will not allow any further moisture loss. Apart from introducing potential contaminants, the application of post-process ingredients has little influence on the safety of most dried or fermented meats with respect to *C. botulinum* as these products are considered essentially stable by the time such ingredients are applied.

The safety of these types of products is perhaps still more an art/craft than a well understood science and it is critical in such circumstances to avoid the temptation to modify the fundamental process technology without a full knowledge of the implications this may have on the safety and stability of the product. Increasing drying temperatures or reducing drying times to speed up production may appear to be practical solutions to commercial needs but ultimately this may lead to compromised product safety. If processors do not fully understand the boundaries between safe and unsafe processes, clearly they will not know when their product development/modification activities may exceed any safety margin. Therefore, any such developments need to be taken slowly and after careful consideration of the impact they may have on product safety, particularly in relation to the hazard of *C. botulinum*.

PASTEURISED, ACIDIFIED, AMBIENT-STORED FRUIT AND VEGETABLES

The use of acidity and low pH is a feature protecting a large number of products against the growth of *C. botulinum*. Products in this category range from pickles and sauces which may or may not be heat-processed to heated fruit relying on natural acidity and pasteurised, acidified vegetables. Such products have been implicated in several outbreaks of botulism (Odlaug and Pflug, 1978), usually due to inadequate acidity of the product or pH elevation caused by growth of spoilage microorganisms. The products are sold as ready-to-eat and may be used to accompany other dishes, as in the use of pickles, relishes and sauces, or they may be consumed on their own as fruit and desserts, e.g. canned pineapple, peaches, etc. The products are consumed by all sectors of society and are usually sold in cans, glass jars or bottles.

Description of process

The manufacturing process for these products varies significantly between products but the main distinction is whether a heat process is applied or not. Most fruit and acidified vegetables receive a heat process during their manufacture, whereas many pickles and sauces may not be heat processed (Figure 4.11). The raw materials and their combinations in products vary markedly and have a significant influence on the level of contamination that the process and product are exposed to. Raw materials include fruit, vegetables, dried herbs and spices, organic acids, salt, water and sometimes preservatives such as benzoate or sorbate. The raw material fruit and vegetables are usually washed to remove soil and dust and some may then be peeled and/or cut/sliced, as necessary. Products such as fruit and vegetables may be acidified with organic acids, usually acetic, lactic or citric acids and then filled into cans or bottles prior to heat processing. Some vegetables are fermented to produce speciality products, e.g. sauerkraut. Pickles and sauces usually consist of finely chopped or blended vegetables mixed with herbs and spices, salt, organic acid (usually acetic acid) and often sucrose. Many of these ingredients can contribute significant spore loading to the product. Most are not cooked but, if heat processed, this is usually performed in vessels prior to hot filling.

All products in this category have pH values of 4.6 or below and, if cooked, receive heat processing well below the $F_0 3$ process required for canned low-acid products. Heat processing temperatures for acid products frequently do not exceed 100°C. The products are sold at ambient temperature with shelf lives varying from several weeks for some

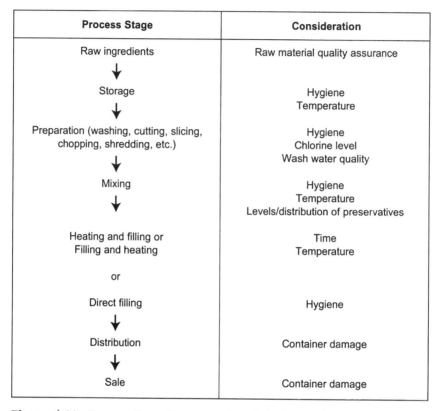

Process Stage	Consideration
Raw ingredients	Raw material quality assurance
↓	
Storage	Hygiene Temperature
↓	
Preparation (washing, cutting, slicing, chopping, shredding, etc.)	Hygiene Chlorine level Wash water quality
↓	
Mixing	Hygiene Temperature Levels/distribution of preservatives
↓	
Heating and filling or Filling and heating	Time Temperature
or	
Direct filling	Hygiene
↓	
Distribution	Container damage
↓	
Sale	Container damage

Figure 4.11 Process flow diagram and technical considerations for typical ambient pH controlled products.

pickles and sauces to over a year for canned/bottled fruit and acidified vegetables.

Raw material issues and control

Clearly, the raw materials used in the production of these products will have a significant influence on the frequency and level of contamination in the product. Unless soil contaminated dropped fruit is used, fruit grown on trees is likely to be contaminated only on rare occasions and at very low levels with spores of *C. botulinum* spread via airborne dust particles. Soft fruit and ground fruit, such as strawberries and raspberries, and most vegetables will be contaminated at a much higher frequency and levels.

Surveys of the incidence of *C. botulinum* in vegetables have been reviewed by Dodds (1993b) and Notermans (1993) and demonstrate the occasional occurrence of spores in many types of vegetables. Although washing of fruit and vegetables will have an impact on reducing levels, such reduction will only be small, achieving perhaps a reduction of one order of magnitude. Peeling, however, can significantly reduce the incidence and levels of contaminating spores and, combined with washing of the peeled fruit or vegetable, may reduce the incidence by several orders of magnitude. Very little data are available on the impact of such practices on microbial loading but it is evident that as most contaminants occur on the surface of the raw material, any process that removes the surface hygienically will reduce the loading significantly. This, of course, will be dependent on the nature of the peeling process and degree of contamination spread by the peeling machine and the wash water used to remove residual peelings. Indeed, if carried out under inadequate standards of hygiene, it is feasible that the peeling process could actually increase levels of contamination and spread them over entire production batches.

For many raw materials used in these products, such as herbs and spices, it may not be possible to effectively reduce contamination levels once received. Dry herbs and spices can carry high microbial loads of general microflora which often include many spore-bearing bacteria, up to 10^5/g. It is important, therefore, to operate effective quality assurance systems designed to select the most appropriate quality for the products manufactured.

It is necessary with any process, whether heat-based or reliant only on inhibitory factors present, to prevent excessive levels of microbial contamination entering the system. Organic acids, for example, can be ineffective in the presence of high levels of particular microorganisms (International Commission on the Microbiological Specifications for Foods, 1980). In processes using raw vegetables and herbs and spices, microbial loads may be difficult to control but, wherever possible, good agricultural practices and adherence to hygienic processing during peeling and washing will make a significant difference to microbial loads entering the process. It may also be useful to monitor the spore loading of the raw materials as excessively high loads may indicate poor agricultural/ harvesting practices and trigger remedial action to improve control at the raw material stage. This may be particularly relevant for herbs and spices where spore loading can be extremely high.

Process issues and control

Many raw ingredients used for these products are stored prior to processing. It may be easy for processors of products receiving a heat process and/or where acidification is used to stabilise the product to believe that these stages of processing will control any contaminants and therefore control of the preceding stages is less important. This would be a dangerous assumption as poor control of raw material due to storage for excessive times or poor temperature or humidity control may allow contaminants to proliferate and compromise the stability and potential safety of the final product.

As the primary hazard with acid-controlled products is the presence of organisms capable of growth and pH elevation in the final product, allowing increases in levels of bacterial spore-formers or moulds in raw materials may compromise safety of the final product with respect to *C. botulinum*. Therefore, raw fruit and vegetables should be stored only for short periods and, ideally, under refrigeration in dry conditions to preclude the growth of microbial contaminants.

Equipment used to peel or wash raw materials should be regularly and thoroughly cleaned to prevent colonisation by, and growth of contaminants and the potential for extensive build-up of populations that can cross-contaminate to subsequent batches of product. Wash water should be changed regularly and chlorine dosing should be used to prevent extensive raw material cross-contamination. Levels of 5–50 ppm or more available chlorine are commonly employed. Chlorine will be rapidly depleted by the high organic loads in the wash water and an important critical control point requiring monitoring is the free chlorine levels in the wash water. The use of continuous chlorine dosing systems can be a considerable practical help.

Many raw, prepared vegetables and most materials for pickles and sauces are mixed in large vessels and filled directly into containers. Cleaning and hygiene of such equipment is again critical to prevent contamination building up. Equipment such as slicers and blenders used to finely chop and mix ingredients can be very difficult to clean effectively and it is essential that such areas do not become foci for build-up of contaminants.

Where products are artificially acidified, the pH and concentration of acid added to the product are critical to its subsequent safety. It is essential for all products relying on acid preservation to ensure equilibration of the acid throughout the product. The larger the raw material pieces the longer and harder it will be to achieve acid and pH equilibration. Studies to validate the times and conditions required to achieve pH and acid equilibration are absolutely essential in the development of these products. Such studies must take account of the size of the fruit and vegetable pieces and such factors must be built into subsequent process control schedules. Using thicker pieces of vegetable than has been demonstrated in validation studies as achieving effective equilibration of pH may result in an unsafe product, with some areas of the material lacking sufficient acidity to inhibit growth. In addition, account may also need to be taken of other factors such as temperature which may affect the rate of equilibration. It may also be important to ensure pH drift does not occur during the life of the product due to slow neutralisation of acids or protein breakdown. Such studies must be undertaken as part of the process validation and should be considered using products undergoing the same heat process, if used, and of the same formulation as the intended final products. Where formulation or processing changes occur, it is necessary to re-validate the safety of the product.

As all of the products considered in this category are controlled to pH 4.6 or below, providing the pH remains at or below this level then the product will remain safe with regard to *C. botulinum*. The greatest hazard in relation to these products is growth of contaminating microorganisms capable of elevating pH as a consequence of growth. Although usually unable to significantly affect the pH of the bulk of the product, growth of mould and some bacteria (especially *Bacillus* spp.) can cause increases in local pH and such conditions have indeed resulted in growth of *C. botulinum* in acidified products as spores of *C. botulinum* can survive for extremely long periods at low pH (Odlaug and Pflug, 1977).

It is therefore normal for naturally acidic or acidified, ambient-stored products to be subject to some form of further preservation mechanism in addition to pH control to prevent growth and spoilage by microorganisms. Conditions usually employed include pasteurisation or addition of other preservative agents.

The temperature used to destroy spoilage microorganisms varies depending on the pH of the product. At low pH (< 3.5) few bacteria capable of elevating pH can grow. Temperatures used to destroy these organisms traditionally achieve 70–80°C for several minutes. Often the process employed may also need to achieve cooking of the ingredients in order to effect organoleptic changes and, thus, the heat process may be significantly longer than that required to achieve microbial stability. Clearly, it is essential to recognise that excessive contamination levels in raw ingredients or the presence of mould spores may require higher temperatures, possibly even exceeding 100°C to achieve stability.

At pH values above about 3.6, microorganisms with greater heat resistance may survive and, if present in the final product, may be capable of growth and pH elevation. Organisms including some *Bacillus* species, are capable of growth at low pH and some species may cause local pH elevation during growth, e.g. *Bacillus licheniformis* (Montville, 1982). Some strains of *C. butyricum* have been reportedly associated with botulism (Aureli *et al.*, 1986), these toxigenic strains were capable of growth at pH 5.2 but not 5.0 (Morton *et al.*, 1990). The non-toxigenic strains tested grew at pH 4.2. Butyric anaerobes are particularly common in tomatoes and other fruit and it is generally recommended that products containing tomatoes and those where butyric anaerobes may be present should have slightly higher heat processes applied to destroy these contaminants.

Guidance given by the National Food Processors Association (1968) in the USA specified that tomato-based products with pH values between 4.0 and 4.3 should be heated to a temperature of 93.3°C for five minutes or an equivalent process. Products at a pH of 4.3 and above (but presumably <pH 4.7) are given a heat treatment of 93.3°C for 10 minutes. Such processes should also make the product stable with respect to other microorganisms capable of elevating the pH, e.g. moulds, providing contamination of raw materials is not excessive. Where tomato products or other ingredients potentially containing high levels of butyric anaerobes are not used as part of the formulation and where the pH is between 3.7 and 4.2, milder heat processes are often applied, e.g. 85°C for 5 minutes or 95°C for 30 seconds (Gaze, 1992). The principle behind most of these processes is to destroy contaminants present in the raw ingredients that may be capable of growth at the final product pH under ambient storage. In many cases these processes also provide security with regard to general product spoilage.

As temperature is so important it is essential to ensure the heat process applied to filled containers achieves the time and temperature required. This is done using heat penetration studies of processes, which will clearly vary depending on the type of product and container, together with the heating equipment being used.

As for any cooked product the heat penetration studies should ensure the heat process is capable of achieving the correct cook under worst case conditions. This is usually achieved using the coldest and largest particle size of raw product placed in the coolest part of the heating equipment and exposed for the minimum duration at the lowest process temperature.

Products of this nature are often filled into glass containers and then heated using water baths or steam in retorts, although for some, the product may be cooked in large vessels prior to filling. In both circumstances, it is essential to ensure that post-processing contamination is prevented, as mould spores or other microorganisms getting into the product after the heating stage may present a significant risk to the product, although in most cases this will be manifest in overt spoilage rather than pH elevation and growth of *C. botulinum*.

The safety of all of these products relies heavily on the acidity and pH of the product. It is therefore essential to ensure these are controlled accurately. pH measurement is an area often subject to poor control and equipment for monitoring pH must be maintained properly, calibrated regularly and personnel using the equipment must be fully trained in its use and calibration. Fouling of pH electrodes by product residues can lead to inaccurate results and effective cleaning of electrodes between measurements is important. The importance of ensuring that the personnel and equipment involved in pH measurement are trained and maintained respectively cannot be over-stressed. It is often recommended that products with a pH close to 4.6 are formulated to below 4.5 to compensate for inaccuracy in pH measurement and pH homogeneity in the batch.

Many products are prepared in bulk and held for extended periods as production stocks prior to further processing. Due to the seasonality of growing, tomatoes are often puréed, acidified and then heat processed and stored in bulk containers. They are usually acidified to pH 4.5 or less and heat processed according to the standards already described and then hot-filled or aseptically filled into lined bulk tanks. Again, it is essential that these bulk containers are not subject to post-process contamination, in particular with moulds. However, as they are to be reprocessed, it is clearly possible to ensure that growth of spoilage organisms has not occurred in the intervening period by visual, organoleptic or even pH checks prior to further processing.

Final product issues and control

Products in this category are usually assigned extremely long shelf lives. As they are stored under ambient conditions, any contaminants, if capable of growth in the final products, will have ample opportunity to grow and spoil the product. In the majority of cases, if this were to occur, the spoilage would be evident visually or by smell after opening and safety would not be compromised.

It is essential to fully understand the interaction of ingredients and controls in products of this nature during shelf life. Care must be taken to ensure that pH drift does not occur during the life of the product perhaps due to interaction of the acid and protein in complex, protein-rich foods. In most cases, if products were either contaminated after processing or not correctly formulated to be stable, growth of contaminating organisms and concomitant spoilage would occur very quickly. For this reason many processors still employ a procedure of incubation of finished products (or a proportion of them) at elevated temperature (usually 25-30°C) for 1-2 weeks to give added assurance that all controls in the process were effective. Clearly such checks should not be necessary if a proper HACCP based approach to process control is used, but nevertheless such measures can prove their value on occasion.

Growth of contaminants capable of elevating pH represents a significant hazard to these products. In general, these can be precluded by applying a heat process described previously in this text and by preventing post-process contamination by hot-filling or heating in pack. The organisms of concern with respect to pH elevation in acidic foods (pH 4.6-4.0) are a number of moulds e.g. *Aspergillus* spp., *Penicillium* spp. and *Cladosporium* spp. and some *Bacillus* spp., e.g. *B. licheniformis*. Extensive growth of these organisms has been shown to create pH gradients in products, which has allowed concomitant production of toxin by *C. botulinum* inoculated into the product (Odlaug and Pflug, 1979 and Montville, 1982).

Of perhaps equal concern are the various reports of the growth of *C. botulinum* at pH values below 4.6. Raatjes and Smelt (1979) demonstrated that *C. botulinum* could grow and produce toxin at pH values as low as 4.0 in broths in the presence of large amounts of precipitated protein where hydrochloric acid was used as the acidulant, although such circumstances are unlikely to occur in real food systems.

Employing the following approaches will minimise the risks from *C. botulinum* in these products:

- use raw materials with low mould and spore loading
- ensure processing equipment has low microbial contamination loads
- heat process to >93°C for 10 minutes
- ensure low pH (<4.3) throughout the product.

Although pH 4.6 is generally recognised as the level at or below which growth and toxin production by *C. botulinum* would not occur, there are a number of reports showing that it can indeed produce toxin in some circumstances even when the pH remains below 4.6 and, conversely, there are some occasions when it does not grow to produce toxin at pH up to 5.0. However, products properly formulated to pH 4.6 or less have an excellent safety record and this is still recognised as the limit for safety of ambient products with regard to *C. botulinum.*

A number of products in this category, such as many of the ambient stable pickles, are made by cold mixing a variety of ingredients. Under such circumstances it is essential that the stability of the formulation is carefully assessed to be capable of controlling surviving microorganisms, moulds in particular. For this reason a number of preservation indexes have been developed by various bodies to try to estimate the degree of stability and safety offered by different product formulations. Probably the most widely recognised is the code for the production of microbiologically safe and stable emulsified and non-emulsified sauces containing acetic acid. Developed by the Comité des Industries des Mayonnaises et Sauces Condimentaires de la Communaute Economique Européenne (Anon, 1992), it is often referred to as the CIMSCEE index. This index is designed for acetic acid-based products and is based on studies using pathogenic indicators (*E. coli*) together with spoilage bacteria (lactobacilli) and fungi. The code is based on the work of Tuynenburg Muys (1971) demonstrating that safety and stability can be predicted using pH, acetic acid, salt and sugar levels.

By entering data relating to the product formulation, including pH and the aqueous concentrations of acetic acid, salt, hexose sugar and disaccharides into the following formula, it is possible to estimate the relative stability of a product under ambient storage:

Σ 15.75 (1 – α) (total acetic acid %*) + 3.08 (salt %*) + (hexose sugar %*) + 0.5 (disaccharide %*)

All items marked with an asterisk (*) are expressed as percentages calculated by dividing the weight of the individual component by the weight of the total aqueous solution, i.e. water plus acetic acid plus salt plus sugars and then multiplying by 100. (1 – α) is the proportion of total acetic acid that is undissociated at the pH of the product, i.e. (1 – α) = 0.998 at pH 2.0, 0.983 at pH 3.0, 0.851 at pH 4.0 and 0.644 at pH 4.5.

The formula takes into account the relative dissociation of acetic acid at the pH of the product and if the result from applying the formula exceeds 63 then spoilage should not occur, whereas at values less than 63 spoilage is likely to occur. Due to variations in pH, it is best practice to ensure products exceed 63 by a large margin otherwise stability may not be consistently achieved. A slightly amended formula is used to predict safety with regard to pathogens (Anon, 1992).

The formulae are very useful when considering acetic acid-based preserves but are not appropriate when other organic acids are used. However, when the index indicates lack of either safety or stability, other challenge tests are suggested to establish product stability. In the context of *C. botulinum*, it is important to recognise that the safety index was not specifically designed for this organism but, as this organism will not grow at the low pH levels specified for use with the code (< 4.5) unless the pH becomes elevated through the growth/metabolism of other organisms, predictions demonstrating stability (with the stability index) with regard to bacteria and moulds will ensure that such spoilage and hence pH elevation does not occur.

As most of these products are given long shelf lives in pack, it is important to recognise that many may be susceptible to growth and spoilage by contaminants entering after the consumer breaches the product seal and uses the product for the first time and on subsequent occasions. Because of this, many of these products have additional advice on the pack label to use the product within a certain time period or to store it refrigerated after opening. Although such instructions do appear on many products of this nature it is possible that many consumers do not heed them due to the apparent ambient stability of the products. Clearly, the processor must take account of this and wherever possible make instructions large, clear and unambiguous.

PROCESSED CHEESE

Processed cheese and processed cheese spreads have a relatively good safety record. An outbreak of type A botulism occurred in 1974 in Argentina associated with commercial cheese spread with onions; there were six cases and three deaths (Briozzo *et al.,* 1983), but outbreaks in these products occur very rarely. In the last two decades there has been much research regarding the control of *C. botulinum* to try to understand more clearly the factors contributing to the safety of these products. Processed cheese and cheese spreads are manufactured throughout the world and are predominantly displayed and stored under ambient conditions, although in the UK most processed cheese is sold under chilled conditions. Although most of the products sold have similar basic formulations, a number also have fish, meat or vegetable additions to give added appeal to the primary consumers, children. The products are most often used in sandwiches and for light snacks and, because of the nature of the products, they may be made with lower grades of cheese as it is to be mixed and processed with other ingredients.

Description of process

The basic product is manufactured by heating cheese with salt and emulsifiers (usually a mixture of di- and polyphosphates) and then hot-filling while molten into tubes, sachets or other containers (Figure 4.12). These are sealed and then sold under either chilled or ambient conditions with a shelf life of several weeks or months. If other ingredients are added, this is done at the heating stage and as they may include skimmed milk powder, whey powder, herbs, spices or pieces of meat, shellfish or vegetables, introduction of microbial contaminants is clearly possible. Some formulations contain the preservative nisin which is variously reported to offer some protection against the growth of *C. botulinum*.

The heating process usually achieves temperatures in excess of 80°C but the process can vary from 80°C to 90°C for several minutes or more. The key safety controlling factors in processed cheese and cheese spreads are the availability of water (water activity) and pH. However, accurate measurement of water activity in products of such high fat content is difficult and should not be relied upon where the water activity distinguishing safe from unsafe formulations may be only a small amount. In most cases, the growth controlling factor is expressed as the salt content in the water phase of the product and, therefore, the parameters of note are the salt level and the percentage moisture. Aqueous salt content in these products can vary from 2 to > 5% and the pH can equally vary from

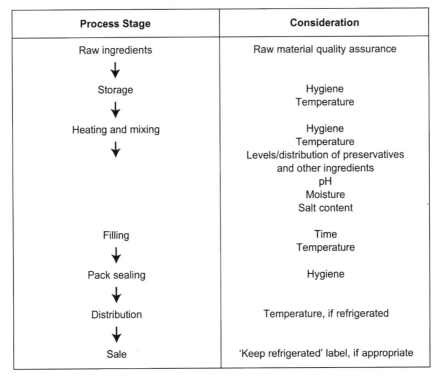

Process Stage	Consideration
Raw ingredients ↓	Raw material quality assurance
Storage ↓	Hygiene Temperature
Heating and mixing ↓	Hygiene Temperature Levels/distribution of preservatives and other ingredients pH Moisture Salt content
Filling ↓	Time Temperature
Pack sealing ↓	Hygiene
Distribution ↓	Temperature, if refrigerated
Sale	'Keep refrigerated' label, if appropriate

Figure 4.12 Process flow diagram and technical considerations for typical processed cheese.

< 4.5 to 6. The pH is significantly affected by the type of cheese used and the addition of organic acids.

Raw material issues and control

The primary raw material used in these products is, naturally, cheese. Cheese for processing may be derived from a number of sources and is usually blended to achieve the correct formulation. Because of the further processing involved in the manufacture of these products, it is possible to utilise cheese of lesser quality than that used for direct sale. The flavour of any particular variety of cheese is an essential component in the selection of that cheese. Processed cheese raw materials can include off-cuts and cheese downgraded from the manufacture of standard varieties. It is therefore possible, but not always the case, that the raw material cheese may harbour a higher microflora than standard products. As such it is important to ensure that appropriate monitoring programmes are in place to assess the microbiological quality of the incoming cheese batches. It may be a common, if misguided assumption that as the process involves a further heating stage the microbial quality of the raw ingredient is of no consequence. However, the temperatures achieved in the process for the manufacture of processed cheese are barely capable of destroying non-proteolytic strains of *C. botulinum* and will have no effect on proteolytic strains. Therefore, poor quality cheese with a high loading of spores may, in fact, compromise the preservation system used for the final product formulation to maintain safety during storage and sale of the finished product. *C. botulinum* is likely to be only an infrequent contaminant of the raw ingredient cheese and it is normal to monitor the cheese for total levels of spore-forming bacteria, primarily because of concerns over the development of gas forming bacteria contributing to spoilage rather than as any indication of the presence of *C. botulinum*. Nevertheless, such tests can provide confidence in the integrity of the raw material supply and help keep spore levels low for the subsequent process.

Of course, cheese is not the only raw material used in these processes. Many formulations include the addition of herbs and spices and some even have meat and/or vegetable additions.

Herbs and spices are usually added as extracts, but clearly, if dried natural materials are used these may contribute significantly to the spore loading of the final product and such raw materials should be monitored to confirm a consistent good microbiological quality. Sourcing from reputable supply sites or using materials that have been microbiologically selected or even processed in a manner to reduce contamination, such as washing before drying, can be an important means for achieving minimal microbial (including spore) loading in the product.

The microbiological integrity of the other ingredients, e.g. ham, prawns, vegetables, must also be considered as these may need to be specifically formulated to control the hazard of *C. botulinum* once incorporated into the cheese product. Considerations such as pH equilibration and moisture content will be critical and this will be particularly so if the piece size of the ingredient is large. It is interesting to note that the one outbreak of botulism associated with processed cheese spread involved a product containing onion (Briozzo *et al.*, 1983) and, therefore, consideration of control of the additives is important.

In addition to ensuring their stability, the potential for the material to introduce spores of *C. botulinum* into the cheese product must also be assessed. Both shellfish and pork are known to be occasionally contaminated with spores of the organism and while there is little that can be done to prevent this, it is important to ensure that any contaminants entering the cheese manufacturing plant are not allowed to increase through ineffective storage conditions. It is good practice to store such raw ingredients under frozen conditions, but it is also common practice to use canned, ambient-stable ingredients.

Process issues and control

Processed cheese spreads usually contain higher levels of moisture and lower solids than processed cheeses and consequently the potential for growth of *C. botulinum* is greater.

The process for the manufacture of processed cheese or processed cheese spreads is fairly simple, involving heating, mixing and a filling operation.

Cheese is heated in a steam-jacketed vessel until molten and other required ingredients are added. Ingredients vary with different processed cheese and cheese spreads but variously include water, skimmed milk powder, whey powder, sodium chloride, emulsifier (di- or polyphosphate or citrate), organic acids (lactic or citric), flavourings (including herbs and spice extracts) and colourings. Some varieties may also have processed raw ingredients added such as pieces of meat, e.g. ham; shellfish, e.g. shrimp; or vegetable, e.g. onion, pepper.

The cheese mixture is usually heated together with the other ingredients to temperatures exceeding 80°C while mixing continuously. The mixture is held at these temperatures for several minutes and then hot-filled into containers prior to sealing.

The maintenance of all equipment in good hygienic condition is an important consideration to ensure microbial contaminants do not build up in the processing equipment. Although unlikely to be an important source of *C. botulinum,* good hygienic standards nevertheless need to be maintained effectively.

Of great importance will be the presence and distribution of the preservative factors in the production mixture. It is essential that the mixing process ensures effective distribution of the ingredients particularly those that will have antimicrobial properties. However, as effective mixing is equally important in achieving product quality characteristics such as flavour and physical attributes, e.g. spreadability, such factors are usually under good control.

Although processed cheese and cheese spreads have a heat process applied, this is usually designed to facilitate effective mixing of ingredients and process flow characteristics for filling, etc. rather than to ensure the destruction of high numbers of contaminating microorganisms. Processes do vary quite widely, from 80°C for a few minutes to in excess of 90°C for 10 minutes or more. In some cases, therefore, the heat process itself may be sufficient to destroy or certainly significantly reduce spores of non-proteolytic strains of *C. botulinum*. However, these temperatures will have no effect on spores of proteolytic strains.

If the product is to be stored under ambient conditions, then the heat process applied will have little influence on the subsequent safety of the process as this must be derived purely from the formulation of the product which must inhibit growth of proteolytic strains of *C. botulinum*. The heat process would however achieve destruction of contaminating moulds, which may be important in terms of shelf life stability.

If the product is intended to be stored under chilled conditions then the formulation may be less preserved than ambient products and greater reliance will be placed on the heat process. If this is the case, then the heat process must be capable of reducing levels of any contaminating non-proteolytic strains of *C. botulinum* with temperatures of at least 90°C for 10 minutes being the recommended minimum process for achieving a significant reduction of non-proteolytic strains. In addition, the potential for post-process contamination in the factory must be minimised as contaminants entering here may be capable of growth in the final product formulation. It is also important to recognise that as these products are afforded long shelf lives, much of which may be in the consumer's own refrigerator, contamination and growth can equally be of concern after packs are opened. Under such circumstances the product should either be formulated to control growth of strains of non-proteolytic *C. botulinum* or have the post-opening shelf life restricted with a clear message on-pack to the consumer to 'Keep refrigerated' and 'Consume within' a defined period after opening.

In practice, most formulations of processed cheese and cheese spreads are capable of controlling growth of non-proteolytic strains of *C. botulinum*, i.e. water activity 0.97 or less and/or pH 5 or less, and the greater concern with these products remains with those stored under ambient conditions or where those destined to be chilled are temperature abused, for which proteolytic strains of *C. botulinum* are the hazard.

Final product issues and control

The safety of processed cheese and processed cheese spreads has been the subject of much research over the last two decades. Evidence exists to show that, if inadequately formulated, the potential for growth and toxin formation by *C. botulinum* is very real indeed.

Kautter *et al.* (1979) inoculated five ambient, retail, pasteurised, processed cheese spreads with a mixture of *C. botulinum* types A and B at a level of 2.4×10^4 heat shocked spores/jar and incubated them at 35°C for up to six months. Two of the five types of spread examined allowed toxin formation after 50 and 83 days, respectively (Table 4.44). One of these cheese spreads was also inoculated with a lower level of spores (460 spores/jar) to determine the effect of spore loading. Botulinum toxin was found after 149 days' storage in one of the jars.

Table 4.44 Toxin formation by *C. botulinum* types A and B inoculated into five commercial brands of cheese spread and incubated for six months at 35°C, adapted from Kautter *et al.* (1979)

Cheese spread	pH	Water activity	Toxin formation
Cheese and bacon	5.70–5.73	0.936–0.941	Yes (50 days)
Limburger	6.26–6.32	0.952–0.953	Yes (83 days)
Cheez Whiz	5.81–5.86	0.932	No
Old English	5.52–5.57	0.930–0.936	No
Roka Blue	5.05–5.09	0.944–0.945	No

Kautter *et al.* (1981) reported results from similar challenge test studies on commercial cheeses including what they referred to as imitation cheese (products containing non-milk fat ingredients and oils). In this series of studies with *C. botulinum* types A and B, inoculated at 800–1930 heat-shocked spores/cheese and incubated at 26°C, only one product became toxic after 26–28 days (Table 4.45). This imitation mozzarella had a high water activity (0.973) and, also, a relatively high pH (5.86).

It is important to recognise that these studies were conducted with processed cheese where the inoculum had been introduced into final product packs. It has been suggested by other researchers that, due to the nature of processed cheese and cheese spreads, this procedure may result in localised pockets of higher moisture and higher levels of the organism where the inoculum has been introduced in an aqueous suspension. Tanaka (1982) inoculated spores of *C. botulinum* types A and B (1000

Table 4.45 Growth and toxin production by C. botulinum types A and B (800–1930 spores/cheese), inoculated into various imitation cheeses and stored at 26°C, adapted from Kauter et al. (1981)

Cheese spread	pH	Water activity	Toxin formation	Day toxin tested*
American processed cheese substitute	5.61	0.952	No	154
American processed cheese substitute	5.6	0.951	No	260
Imitation cheddar cheese	5.61	0.960	No	144
Shredded mozzarella substitute	5.56	0.961	No	233
Mozzarella cheese substitute	5.86	0.973	Yes	26–28
Imitation low-moisture mozzarella	5.53	0.961	No	14
Imitation low-moisture mozzarella	6.14	0.958	No	278
Imitation pasteurised processed cheese	5.99	0.942	No	272
Imitation pasteurised processed cheese spread	5.9	0.944	No	267
Pasteurised processed cheese-filled food	5.72	0.946	No	289
Provolone cheese substitute	5.55	0.957	No	252

*Day toxin tested varied as many products spoiled during the trial.

spores/g) into a variety of pasteurised processed cheese spreads using a 'hot' process (inoculum introduced into molten product during the process) and a 'cold' process (inoculum introduced into the final product). Product was stored at 30°C and 35°C for periods up to 49 weeks and tested for toxin. Product formulations found to be safe, i.e. no toxin produced, when inoculated by the 'hot' process were found to be capable of supporting growth and toxin production when inoculated using the 'cold' process. Studies on the localisation of spores confirmed that the concentration of the inoculum and the dilution effect of added water on the local salt concentration were likely causes of the differences. Therefore, it is important to interpret results using 'cold' inoculation with care and perhaps, as worst case situations.

Other studies have been carried out on the growth and toxin formation by *C. botulinum* by inoculating the mixture during processing and these studies are likely to better reflect the conditions that the organism is exposed to during the production of these cheeses. Probably the most definitive early work in this area was published by Tanaka and colleagues, who produced models to predict the growth of *C. botulinum* in the products. Tanaka *et al.* (1986) produced a variety of processed cheese spreads to investigate the effect of different factors on botulinum growth and toxin production in the different formulations. They found that water activity was not a reliable measure of control of *C. botulinum* due to the difficulties in measurement and the small dynamic range of different formulations. Using moisture, salt and pH, they produced boundary plots showing conditions where the organism would and would not produce toxin. In this way it was possible to predict the safety of processed cheese spread formulations with moisture contents of 51%, 52%, 54%, 56%, 58% and 60%. Clearly, they found that as moisture content increased, the required pH and combined salt content (sodium chloride and disodium phosphate) needed to be increased to inhibit the growth and toxin production by *C. botulinum*. Some estimates from the graphs of Tanaka *et al.* (1986) are given for different moisture content processed cheese spreads in Table 4.46.

It is interesting to note that Tanaka *et al.* (1986) found that sodium chloride and disodium phosphate inhibited *C. botulinum* to the same extent and their combined concentrations could be used to estimate the inhibitory effect on *C. botulinum*. They also reported increased inhibition with lactic acid, mostly attributable to the associated pH decrease, although they did report a slight additional inhibitory effect beyond that of pH alone.

One possible limitation of the Tanaka models is that they appear to

Table 4.46 Conditions of salt content (sodium chloride and disodium phosphate) required to inhibit growth of *C. botulinum* at 30°C for 42 weeks with fixed pH and varying moisture content, adapted from Tanaka *et al.* (1986)

Moisture content (%)	pH	Salt content (%) (approximate)*
51	5.8	4
52	5.8	4.2
54	5.8	4.6
56	5.8	5
58	5.8	5.4
60	5.8	5.8

* Estimated from graphical data.
Note: salt content is not aqueous salt.

underestimate the protection afforded by lower pH processed cheese spreads as their emphasis had been on moisture with challenge studies conducted at fairly high pH values (Ter Steeg and Cuppers, 1995).

The latest and perhaps most comprehensive study on processed cheese and cheese spreads was the development of a mathematical model to predict growth of *C. botulinum* by Ter Steeg *et al.* (1995) and Ter Steeg and Cuppers (1995). This group examined the effect of temperature, moisture, pH, phosphate and citrate emulsifiers and lactic acid presence and levels on the growth of and toxin production by *C. botulinum* types A and B.

The latter authors used lower pH values (5.45–5.9) in their studies and found that the key controlling factors were pH, temperature, sodium chloride and emulsifying salts. Temperature was shown to significantly affect growth, with spore outgrowth occurring at 18°C after only three months under conditions of a_w 0.966 and pH 5.9 whereas at 25°C this occurred within one week under the same conditions. They found that growth was inhibited in processed cheese at a pH of 5.55 and a_w 0.95 when incubated at 25°C for three months. Emulsifier was reported to be an important determinant of safety, with citrate offering less protection than polyphosphate. In products where the fat on dry matter content deviated from 50%, they reported moisture level to be a poor determinant of product safety with better predictions being possible using pH, sodium chloride, emulsifying salts and lactic acid levels (the latter three all in the aqueous phase). Spore loading did not affect ultimate toxin production but merely affected the time to toxin formation with low levels requiring greater time before toxin was detectable.

Somers and Taylor (1987) studied the effect of nisin on the production of toxin by *C. botulinum* in processed cheese spreads. Using a spore inoculum of 1000 spores/g (types A and B) and then incubating cheese spreads at 30°C for up to 48 weeks, they found that nisin could prevent toxin formation in reduced sodium chloride formulations. Although batches were somewhat variable in moisture level and pH (5.7–6.2), it was observed that processed cheese spreads formulated to a target of 54% moisture and with 2.5% added disodium phosphate and 2% added sodium chloride in the absence of nisin prevented toxin production. When sodium chloride was removed from this formulation, one sample became toxic within the 48 week incubation at 30°C (Table 4.47). The addition of 12.5 ppm nisin in the same formulation, i.e. without sodium chloride, prevented toxin formation. In high moisture processed cheese spreads (target 58% moisture), the presence of 2.5% disodium phosphate and 2% sodium chloride did not prevent toxin production. Toxin was produced in 79 samples of this formulation as early as four weeks. Nisin (100 ppm) added to this high moisture formulation significantly reduced the number of toxic samples (only one found) but did not prevent it completely. In lower moisture product (target 54%) with reduced phosphate (2%) and no salt, all 80 samples developed toxin (first samples toxic after four weeks). The addition of 50 ppm nisin to this formulation reduced toxin production to only five samples while 100 ppm prevented it completely. These results, while needing to be interpreted carefully because of the variations in pH and moisture level of different batches, clearly show the benefits of the use of nisin in lower salt formulations of processed cheese spreads.

Karahadian *et al.* (1985) studied the effect of reduced sodium and different emulsifiers on growth of and toxin production by *C. botulinum* types A, B and E inoculated into molten processed cheese foods and spreads at 1 spore/g and 1000 spores/g after extended storage at 30°C. They found that safe formulations could be produced with low sodium, providing pH was maintained at low levels (5.26 or less) using deltagluconolactone. It was also observed that the use of potassium emulsifiers in place of sodium phosphates provided less protection and disodium phosphate gave greater protection in comparison to citrate emulsifier.

In formulating processed cheese and processed cheese spreads, it is critical to ensure the factors selected and used to inhibit the growth of *C. botulinum* are clearly understood and well controlled both for added ingredients and the final product. The data published by Tanaka *et al.* (1986) and Ter Steeg and Cuppers (1995) can be helpful when assessing the safety of these ambient stable products.

Table 4.47 Toxin production by *C. botulinum* types A and B at c. 10^3 spores/g in processed cheese spreads and the effect of nisin, adapted from Somers and Taylor (1987)

Batch	Moisture (%)	Disodium phosphate (%)	Sodium chloride (%)*	Nisin (ppm)	Number of samples toxic†
3	52	2.5	0	12.5	0
1	52.6	2.5	2	0	0
2	53.2	2.5	0	0	1 (24 weeks)
26	55.9	2.5	2	0	79 (4 weeks)
28	57.2	2.5	2	100	1 (24 weeks)
7	51.6	2	0	0	80 (4 weeks)
9	53.8	2	0	50	5 (8 weeks)
10	54	2	0	100	0

Incubated at 30°C for up to 48 weeks. pH 5.7–6.2.
* Not aqueous salt.
† 90 products sampled per batch.

In any case, the application of HACCP-based systems to determine the likely hazard from all raw materials, the effect of processes used and the effect of final product formulation and shelf life conditions remains essential for assuring the consistent safety of these products.

LOW-ACID CANNED FOODS

Canned foods are synonymous with the risk of botulism. This is unsurprising given the large number of outbreaks of botulism that have occurred over the years associated with canned foods. However, most of the outbreaks have been associated with home canning and there have been relatively few large-scale commercially attributable outbreaks of botulism. Most of the home associated outbreaks have involved canned vegetables, but commercial outbreaks have occurred, most notably with canned fish caused by type E strains of *C. botulinum*.

The origins of canned foods can be traced back to the 1800s, the pioneer of this technology being Nicolas Appert. Canning processes revolutionised the storage and distribution of food and today, canned foods represent one of the biggest sectors of the food market. Although somewhat more advanced than two centuries ago the principles of canning defined by Appert remain the same to this day; these being to place the food in a sealed container impenetrable by microorganisms and to apply a heat process to destroy those present inside the container.

Canned foods encompass all commodity sectors from fruit, vegetables, meat and fish to dairy products. They are sold to be consumed cold, after heating or after cooking and are clearly consumed by all sectors of the population.

Description of process

In principle, the canning process is relatively straightforward. Products usually undergo some degree of pre-processing from washing/peeling, slicing or dicing and/or mixing of ingredients for composite foods (Figure 4.13). Ingredients are often pre-cooked during preparation and then either cooled and stored or filled into cans while warm and then immediately retorted. Products in cans are usually pre-heated to ensure that a consistent temperature is achieved prior to the main heat process and then retorted in steam pressurised retorts to the appropriate temperature for the appropriate time.

The design and operation of the retort used can vary significantly depending on the type of containers to be processed. The purpose of this text is not to describe the different types of technology available for canning but to focus on the means whereby the process effectively yields a safe final product. For more detailed texts on canning technology, the reader is referred to the following publications: National Food Processors

Process Stage	Consideration
Raw ingredients	Supplier quality assurance
Storage	Hygiene Temperature
Can integrity	Can/lid quality control checks Supplier quality assurance
Preparation/mixing	Hygiene Temperature
Filling/seaming	Time Temperature Seam checks
Retorting	Time Temperature
Cooling of cans	Hygiene Disinfection of cooling medium
Drying	Hygiene
Conveying and labelling	Can damage Hygiene
Storage and distribution	Can damage Incubation checks
Retail display	Can damage
Consumer	Can damage

Figure 4.13 Process flow diagram and technical considerations for typical canned foods.

Association (1995), CCFRA (1975, 1977, 1997a, b and c) and Department of Health (1994).

Raw material issues and control

Canned food is a description of the process technology rather than the foods themselves and, as a process, it can be employed for any type of food material. As a consequence, the raw ingredients used in the various canned foods can include all food types. Nevertheless, the same principles for achieving safety apply to all ingredients used in canned food.

Firstly, as the heat process applied to these products is usually designed to achieve a specific log reduction in bacterial spore loading, the higher the loading in the raw ingredients, the greater the chances of some surviving in the final product. Ingredients with high spore loading such as herbs and spices, many vegetables and some meat and fish may contribute significantly to the spore loading of the product. Therefore, it is usual to monitor the quality of such raw ingredients as part of a supplier quality assurance programme to ensure that poor quality material with excessively high spore loading is not being received/used. Applied in a trend analysis quality control system, microbiological monitoring of spore loading can be useful in preventing high spore loads entering the system.

Clearly, it is not possible to eliminate spores from the raw ingredients and it must be anticipated that *C. botulinum* will be an occasional contaminant in the raw ingredients. The aim of raw material quality assurance programmes must therefore be to ensure a consistent supply of ingredients with known, controlled levels of spore loading, avoiding wide fluctuations in levels which could ultimately compromise the safety of the final product.

Once received by the manufacturer, the objective must be to ensure that raw ingredients are stored and treated in a manner that precludes proliferation of contaminating microorganisms during the raw material processing stages and prior to retorting.

Raw ingredients should be stored in areas designed to prevent their contamination from sources such as poorly cleaned or maintained equipment or from the environment. Additionally, they should be stored under conditions which preclude organism growth; dry ingredients should be stored in clean, dry storage areas with the aim being to prevent ingress of moisture to an extent that could allow growth. Perishable products like meat and fish should be stored under chilled or frozen conditions and used within the allocated shelf life.

It is essential to recognise that the heat-processing stage in the canning process is not a substitute for ensuring that safe raw ingredients are sourced. The aim of the canning process is to make products ambient-stable and safe, not to 'clean up' unsafe, improperly manufactured or poorly controlled raw ingredients. Supplier quality assurance programmes should be employed to ensure that raw ingredients are produced safely with respect to the potential presence of pathogenic microorganisms and their toxins.

Process issues and control

Most materials used in a canning process undergo some processing prior to the retorting stage. Vegetables will usually be washed and trimmed and are often chopped or sliced prior to adding water or brine and then retorting. Meat may be minced or diced and cooked with spices prior to filling into cans and retorting. For composite foods, such as soups, sauces and prepared meals, the ingredients are combined and cooked in kettles or mixers prior to filling into cans and processing. While many of the products have antimicrobial compounds added to them as part of the ingredients, such as organic acids and salt, these are added as flavourings and seasonings and not as botulinum controlling factors.

As many of the main ingredients are prepared in a mixing or chopping/slicing stage while raw and prior to processing, it is essential to prevent build-up of microbial contamination in/on equipment used due to poor cleaning of equipment and utensils. It is a flawed assumption that just because the product is to receive a high heat process, then control of the stages preceding this process has no effect in relation to the final safety of the product. Poor cleaning and handling practices associated with the raw material may allow proliferation of contaminating spore formers that could lead to excessive spore loads entering the process. This has the effect of increasing the risk of the organism surviving the heat process stage. Every effort must be made to prevent the equipment and handling stages prior to retorting being sources of additional or increased contamination.

Many processes, particularly those for canned meats and fish, involve pre-cooking the product to aid processing, in the case of fish to aid skin and flesh separation. Temperatures and procedures employed can bring the fish/meat well within the growth range of *C. botulinum*, i.e. 30–48°C. In addition, many products are manufactured in factories with no in-process cooling facilities and where the ambient temperature can be extremely high, e.g. 30°C. Ingredients such as fish may thus remain within the growth range of *C. botulinum* for extended periods prior to the retorting stage. Clearly, the potential for growth and elevation of bacterial spore numbers must be considered as part of a hazard analysis based study for any process with preparation stages in which the raw ingredients are held within the optimum growth range for *C. botulinum*. Such concerns are equally important in relation to other toxigenic bacteria.

It cannot be over-emphasised that the retorting stage must not be relied upon to 'clean up' poor-quality raw ingredients or raw ingredients that have received extensive temperature abuse.

After preparation, the product is filled into cans. Fluid products are often pre-heated to achieve a constant temperature prior to filling into cans and more solid and particulate products like fish and meat may be pre-heated once deposited into the can.

After filling the can, the can lid is applied and a double seam is formed which interlocks the lid and the end of the can body in two mechanical rolling operations. The cans then enter the retort.

It is essential that products entering the retort are at the same temperature, as poor temperature distribution in the raw products in cans may result in variable product temperature profiles across the retort batch, and while at best this may result in varied product consistency, e.g. texture, at worst it may result in compromised product safety.

Factors which are important for consideration in ensuring efficacy of the retort process include all of the following:

Fill volume: different quantities of product within a can will require different process times and temperatures to achieve the required F_0 process; therefore effective control of consistent fill volume/weight is critical to safety.

Head space gas volume: the amount of air left in the can may affect the heat penetration to the product and cans with too much air may not be heated sufficiently. In most cases the fill volume/weight gives an indication of the head space volume: the lighter the fill the larger the head space.

Density/viscosity: a more dense/viscous product will require a longer time to raise in temperature due to poorer/slower heat penetration and, where this could be significant, it may be important to ensure that viscosity checks are conducted on products.

Piece/particle size: products with varying piece sizes will require different times to achieve full heat penetration and process; therefore, control of piece size in cutting/slicing/dicing stages may be a critical factor for some products.

Product temperature: starting temperature has already been mentioned and it is critical to ensure that the starting temperature of all units in the ingoing retort batch is consistent to ensure that the subsequent heating profile is also consistent and an effective process is achieved for all units in the batch.

All of the relevant factors must be under tight control to ensure that the heat process applied delivers the predetermined safe cook to all cans in the retort. It is therefore essential that mixed batches with variations in these factors are not processed together.

Further critical elements prior to retorting are can integrity and can seam integrity. Cans are usually received at canneries as intact (one- or two-piece) cans but have, historically, also been received as flat-body packs, which are reformed on site. The reforming process for cans needs to be carefully controlled as reforming can result in physical damage leading to seam weakness or even tears and pin holes in the can. Several types of can have evolved over the years. Probably the best are those with the least number of seams/welds, i.e. where the can is a single intact body to which a lid is attached. Many cans have a seamed body to which both a bottom lid and a top lid are sealed. In the latter case it is clear that seam faults could occur in three joints whereas in the former there is only one possibility for seam faults to occur.

With the exception of the retorting process, the most critical point in canning is sealing the lid to the can body to make an effective hermetic seal. The seam has to be formed so that the lid and body of the can create a double seal, with each component overlapping a precise amount under a carefully controlled pressure to ensure the hermetic seal is capable of being formed. Factors such as the size of the body hook, seam overlap and seam tightness are critical to ensuring the seal is effective.

In addition, the type of lining compound, which forms part of the hermetic seal at the seam, is also important in ensuring effective seal integrity, particularly as this becomes more fluid during the retorting process. For more detailed information on can seams and seaming operations the reader is referred to texts noted at the beginning of this section.

All of these elements are critical to product safety – the can integrity, the seam integrity and the quality/characteristics of the lining compound. All of these must be subject to extensive process monitoring and control. A significant array of can inspection tests are employed as part of the quality assurance procedures in canning operations. Many of these are conducted on the cans and lids after receipt of a batch and prior to release of the batch into production. Clearly, final can seam integrity can only be assessed during production but must be examined against a strict, pre-designed specification.

Once lidded and seamed, the cans are conveyed into the retort. The time between filling and retorting must be carefully controlled otherwise this may contribute to temperature abuse and microbial proliferation in the product prior to heat processing. If the retort process is designed to operate based on a particular ingoing raw material temperature, excessive delays may result in the ingoing temperature dropping and the subsequent heat process being less effective.

Retort design can vary significantly but the most common types operate on the same principle, that is, as pressurised steam vessels which first evacuate air from the chamber to assist effective steam cooking; this is followed by the application of steam and pressure to temperatures in excess of 115°C. Thermal processes are designed to ensure the can contents receive a minimum time and temperature combination equivalent to 121°C for three minutes, the F_o3 process or $12D$ 'botulinum cook'. This is usually achieved by raising the product temperature to a holding temperature either below 121°C for periods in excess of 3 minutes or to a temperature in excess of 121°C for usually less than three minutes and calculating the delivered F_o value from the 'at temperature' holding phase of the process. In this way the cook has a degree of extra safety built into the process due to the additional heat input derived during the heating up and cooling down phases.

In practice, many of the thermal processes employed for canned, low-acid foods significantly exceed the F_o required to achieve a $12D$ reduction in *C. botulinum*. This is because the process has not only to destroy this potential pathogen but, equally importantly, has to destroy spore-forming bacteria capable of growth and spoilage of the products, the spores of many of which are significantly more heat resistant than *C. botulinum*. As a result, many vegetables are processed to F_o values of 10–15 or more.

Like any cooking process, the efficacy of the retorting stage must be carefully and comprehensively validated prior to commercial operation to ensure that all variables that affect heat penetration and process efficacy are fully understood and properly controlled. The process efficacy study must ensure that the desired F_o is achieved under all worst case conditions. This must therefore include, where applicable, consideration of the following factors:

- cold spots within the retort, determined using heat distribution studies
- lowest ingoing product temperature
- largest piece/particle size in the can
- largest volume/weight of product in the can
- lowest operating temperature of the retort
- most viscous/dense product
- largest headspace in the can.

Processes must be established taking these 'worst case' factors into account. Calibrated temperature monitoring equipment such as thermocouples linked to chart recorders should be used to monitor and record process temperatures. These studies should be repeated on several occasions for each different type of product and for different product sizes to establish the retort processing conditions that are required to achieve the appropriate in-pack F_o. Equipment is available that incorporates thermocouples and F_o integrators which automatically calculate the F_o achieved from a recorded retort process profile.

Once a process is validated and in operation, validation should be repeated at regular intervals. The process operation itself should be controlled on time and temperature and there should be separate temperature and pressure gauges that are independent of those controlling the process. These will provide an independent indication of the process parameters e.g. temperature/time achieved. All process monitors should be recorded using a chart recorder and the process controls should ideally be linked to an alarm so that any process not achieving the correct temperature is immediately identified and action can be taken, as appropriate. Chart records and reference thermometers should be checked against specification after each process and 'signed off' by designated personnel.

Procedures should also be in place to ensure that a physical indicator of batch heat processing is present to avoid any confusion between batches that have or have not been retorted. Heat indicator tape or ink attached to retort crates are most commonly used as indicators of processing. Such systems are generally unnecessary in larger, more sophisticated manufacturing units where the factory flow is unidirectional and confusion of processed and unprocessed material would not occur. However, many canneries do not operate such product flows and the opportunity to mistake unprocessed product for finished products is a reality.

After retorting, the products may be cooled in the retort with air or water sprays and/or they may pass into cold water baths prior to being dried and conveyed to a packing area. During cooling the can is vulnerable to contamination. The can sealant is highly flexible while warm, and water can be drawn into the can due to contraction of the can contents as they cool down. Because of this, water used to cool cans should be disinfected. Cooling water used may be previously pressurised water from the retort, which clearly will have low microbial loading, but, more frequently, it is potable water treated with chlorine at levels of 1–5 ppm.

Chlorinated water is reported to be highly effective at destroying spores of *C. botulinum* (Ito and Seeger, 1980) and, if carried out effectively, is a good means for decontaminating cooling water (Table 4.48). However, cooling water is often kept for excessive periods and can become fouled with organic debris resulting in poor efficacy of the chlorine. Cooling water must be changed regularly and free chlorine must be present to ensure effective protection against recontamination. In addition, any tanks used for cooling water must be cleaned thoroughly as part of the process equipment hygiene system. A number of canners use simple direct mains supplies of potable water and while this is probably of reasonable quality, it is generally recommended that cooling water used for cooling cans should receive some further disinfection. Cooling water should also be monitored regularly for levels of microbial contamination and records kept.

Cans usually travel on conveyors to packing lines and, as they may be wet after the retorting/cooling stage, many canners include a can-drying stage using hot air to remove water which could otherwise increase the risk of microbial contamination entering the can. In addition, to minimise the potential for contamination from wet conveyor lines, conveyor belts and guides must be kept clean and as dry as possible.

Processed cans should not be allowed to come into contact with equipment, personnel or the environment from the raw side of the factory. This is not always found in practice and use of common areas and equipment for raw and processed materials can lead to significant cross-contamination potential, e.g. placing gloves/aprons used by raw fish handlers onto cooling cans was thought to be a possible factor in the botulism outbreak associated with canned salmon (Stersky *et al.*, 1980).

In a review of 40 canneries in the UK (primarily small manufacturers) it was clear that many of the factors considered to be basic, good canning practice were not being adhered to (MAFF, 1996). While most canneries operated good can seam analysis checks, one canner was conducting no regular seam checks. Several companies had no or inadequate records of heat processing. A number had uncalibrated temperature monitoring

Table 4.48 Effect of disinfectants on spores of *C. botulinum* at 25°C, adapted from Ito and Seeger (1980)

Disinfectant	*C. botulinum* type	Time to achieve reduction (minutes)	Reduction achieved (%)	pH	Concentration (ppm)
Chlorine	A	10.5	99.99	6.5	4.5
	B (proteolytic)	12	99.99	6.5	4.5
	B (non-proteolytic)	5.5	99.99	6.5	4.5
	C	3	99.99	6.5	4.5
	E	6	99.99	6.5	4.5
	F (proteolytic)	8	99.99	6.5	4.5
	F (non-proteolytic)	7	99.99	6.5	4.5
Chlorine dioxide	A	13	99.9	6.7	130
	B	14	99.9	6.5	125
Ozone	A	2	99.9	6.5	6
	B	2	99.9	6.5	5

devices and some had no master temperature indicator devices. Several canneries did not sanitise the cooling water and only 15 of the 40 used process sterilisation indicators to distinguish processed from unprocessed product.

The factors associated with spoilage caused by post-process can handling operations were studied by Put *et al.* (1980). They identified the three most important factors to be the poor condition of the can double seam, the presence of bacteria in cooling water and can abuse due to poor operation or adjustment of the can handling operation. Indeed, defective closures of lids was the cause of a type E botulism outbreak involving canned tuna in Michigan in 1963 (Stersky *et al.*, 1980).

Strict control of the canning operation is clearly important in ensuring the safety of these products. Detailed recommendations on such control is given in a most useful reference document from the UK Department of Health (Department of Health, 1994).

Final product issues and control

Low-acid canned foods are given long shelf lives ranging from 1 to 5 years and the shelf life is usually dictated more by the organoleptic deterioration during life than by microbiological considerations. However, the opportunity for product contamination does not end when the product leaves the manufacturer. Cans are often subject to poor handling practices, many becoming damaged during transportation, sale or even after purchase. In most cases, damaged cans resulting in leakage are self-evident and not used. However, damage to cans occurring at the end seam or side seam, if present, can result in pin-hole fractures which may allow ingress of contamination with little visible signs of leakage. In most of these situations spoilage of the can occurs due to the ingress of contaminating vegetative microorganisms, as it is more likely that a spoilage microorganism will be a post-process contaminant than *C. botulinum*. Nevertheless, handling practices to avoid can damage are extremely important in helping to manage the risk associated with these products. Obviously damaged cans should not be used or offered for sale.

Canned foods are often held for a time in storage by the canner and a number are subjected to high temperature (30–37°C) incubation checks to help assure that batch failures have not occurred. Such practices should not be necessary given that strict process controls should have been adhered to for all the processes. However, many manufacturers and retailers derive some degree of extra assurance from such storage and tests and on occasion these can prove of some, albeit limited, use.

The vulnerability of low-acid, canned foods to contamination via can damage at any stage after manufacturing needs to be understood by all those in the can handling chain. The adherence to controls identified through proper hazard analysis can ensure continued safety of manu-factured canned foods.

GENERIC CONTROL OF *C. BOTULINUM*

Raw material contamination

The quality of the raw materials used in the manufacture of foods can have a significant influence on the microbiological safety of the final product. Even though *C. botulinum* is a spore former and the vast majority of raw materials cannot be assured to be free from the organism, the levels and frequency of its presence in raw materials affect the risk of it surviving or growing in the subsequent final product.

Raw material control is therefore important to ensure microbial levels are not excessive and, indeed, it is also essential, for complex raw materials, that the raw material itself does not present a botulinum risk in terms of its formulation or processing and storage conditions.

As most raw materials are not under the direct control of the final product manufacturer, their microbiological quality and safety are subject to effective production by the raw material supplier. It is therefore important to operate a formal raw material supplier quality assurance programme, including a raw material specification, to ensure systems are in place to control the safety of the raw materials through all stages of the process prior to receipt. This may be difficult to achieve for all raw ingredients as some materials such as herbs and spices may be difficult to trace to source, but every effort must be made to establish and maintain the safety of raw materials. An awareness of the primary source of critical raw materials is important for the purposes of reliable safety assessment.

The following can serve as a useful reminder of the elements that should be in place for effective control of raw materials.

1. Detailed understanding by the raw material user of the production process of the raw material and knowledge of the critical control points or those stages in the process influencing control of C. botulinum

Manufacturers should have an understanding of the processes involved in the production of critical raw materials together with the hazards presented by the raw materials and the factors necessary for controlling safety in the raw material production process. Knowledge of this nature can clearly help in the identification of raw material suppliers who are better placed to provide safe raw materials of high quality on a consistent basis. It is equally important to understand such factors when assessing the effect on final product safety, of changes in raw material formulation or processing conditions that may be introduced for reasons including improved product quality. Indeed, by operating an effective raw material supplier assurance programme based on sound knowledge of the raw material production process and the controls necessary for safety, it may have been possible to avoid the outbreak of botulism (due to toxin formation in the hazelnut conserve) caused by hazelnut yoghurt in the UK.

Raw material suppliers should be expected to ensure the safety of their products by effective operation of hazard analysis based programmes, but a fundamental understanding by the final product manufacturer of the safety requirements for materials to be used in a final product can clearly assist in product safety assurance.

Products relying on preservation factors, e.g. fermented meats, canned cured meats and pH-controlled pickles, may be vulnerable to the growth of *C. botulinum* if excessive levels are present in the raw material due to inadequate control by the raw material supplier. Therefore, as well as ensuring that raw ingredients are not botulism hazards themselves, it is also important to operate supplier quality assurance programmes to help control levels of all spore formers, including *C. botulinum*, as excessive levels may be too great for the subsequent processing or final product formulation to control safely.

2. Audit of the raw material supplier to review process control

The operation of effective raw material supplier quality assurance should involve the operation of a formal auditing programme of, at least, those raw material suppliers whose products are considered to be critical for the safety of the final product. High-risk ingredients, which may be more vulnerable to contamination or have greater inherent botulism risks, should be targeted for more frequent and formal audits. The content and scope of the audit should be clearly defined and must encompass those elements critical for safety.

3. Raw material verification checks

Raw material production processes, operated using the proper application of hazard analysis principles by the raw material producer, should generally preclude the need to conduct extensive microbiological screening of ingredients against complex microbiological specifications.

Indeed, in the case of *C. botulinum*, testing for the organism is rarely conducted, and, certainly, never as part of raw material quality assurance programmes. However, where the levels of contamination may need to be controlled in relation to the subsequent production process, it may be useful to employ some microbiological monitoring of raw materials for other microorganisms indicative of poor production practices. The most commonly applied tests include those for both spore-forming bacteria and sulphite reducing anaerobic bacteria. It is, however, important to recognise that these tests give no indication of the likely presence of *C. botulinum* but, like any indicator of contamination, the presence of high levels may give greater cause for concern in relation to practices that may allow *C. botulinum* and other similar hazards to be present. Applied judiciously, such tests can help in ensuring raw material batches are not of poor quality, and by discussing trends with the raw material supplier it may help improve quality and reduce subsequent levels, thus ensuring the consistent achievement of agreed targets.

4. Agreed specification with the raw material supplier

The safety and quality of the raw material are achieved by a full understanding and the application of the necessary controls required for its production. The specific requirements necessary for the delivery of raw materials of consistent quality and safety must be agreed and understood by both the producer of the ingredient and the purchaser of that material. These are usually detailed in a product specification. It is essential that elements necessary for consistent safety and quality are clearly documented in such specifications agreed and signed by both parties.

5. Conditions of storage and use of the raw material

Raw materials, once received by the purchaser, must be kept under the conditions of storage agreed with the raw material supplier. These conditions must be clearly understood by the receiving manufacturer, as inappropriate storage may lead to inadequate control of *C. botulinum*. For example, any requirement to store ingredients under refrigeration conditions or to use the material within a certain time period must be clearly labelled on the material packaging and understood by the receiving party. Indeed, failure to comply with these requirements has been the cause of outbreaks of botulism in the past, most notably some restaurant associated outbreaks of botulism associated with garlic in oil.

As well as the specified controls of raw ingredients, general precautions to keep them in dry, pest-free storage areas, under effective refrigeration or freezing conditions, if chilled or frozen, and used within the shelf life intended are all important for ensuring safety.

Production incorporates a process to eliminate the hazard

It is common for a number of products for which *C. botulinum* is a hazard to employ processes designed to eliminate the organism. In the majority of these cases, the process employed to achieve this is heating to high temperature. Whether the process is designed to achieve a 'botulinum cook' for non-proteolytic strains equivalent to 90°C for 10 minutes, or for proteolytic strains equivalent to 121°C for three minutes, the same principles apply, namely that all parts of the product are elevated to the desired temperature for the appropriate time and that the subsequent product does not become recontaminated.

Where the heat process is critical to safety of the finished product, process validation studies must be conducted as part of the product development process to identify the exact process conditions necessary to achieve the correct process efficacy for all parts of each product and for each product in the batch. To achieve this, the heat distribution characteristics of the heating equipment must be established and any cold spots identified and eliminated or, if not eliminated, accounted for in the subsequent process validation.

A validation study should be carried out using product of the appropriate final formulation and packaging type. Temperature probes should be inserted into 'worst case' products and the process temperatures and times that ensure an effective cook in these products must be determined. Worst case factors must take account, as appropriate, of the following:

- size and weight of final product
- batch size, i.e. number of products in each batch
- density and viscosity of product
- headspace in product packs, if bottled or canned
- cold spots in cooking equipment
- minimum holding times
- minimum process temperature
- lowest ingoing raw material temperature
- largest piece size of raw ingredients

This will establish the cooking times and temperatures required to achieve the required cook of the product. However, once established, all of the relevant factors detailed above must be subject to process control during routine production and process monitoring.

All heating devices should be monitored during every cooking process to ensure they deliver the correct temperature for the correct time. The cooker should be linked to an automatic alarm so that any deviation from the correct temperature is immediately reported for appropriate remedial action. Temperatures and times actually applied to the cooking chamber and the product should be subject to continuous monitoring and recorded using chart recorders. Such records should be checked following each production batch or at appropriate intervals, and any deviation actioned. The effective control of the cooking operation is critical for products reliant on this for final product stability and safety.

It must also be remembered that the components or formulation of some products may protect the organism from heat processing. This usually occurs only in those products of low water activity or high fat products. Where there is concern regarding the protection afforded by such product characteristics, the process should be validated using challenge test studies in appropriate experimental facilities to demonstrate that the process applied delivers a safe reduction in a known population of the target types of *C. botulinum*.

In addition to applying a correct cooking process, it is also essential to ensure that the product is not subsequently recontaminated during storage, filling or handling, after the heat process. The procedures for ensuring post-process contamination does not occur in canned or bottled foods are well established and include effective seal and seam integrity, disinfection of cooling water, effective cleaning of transfer lines, etc. Also, for UHT products, the controls required in aseptic filling processes are clearly established. Some products, such as dairy desserts destined to be stored under refrigeration conditions, are not subject to in-pack pasteurisation or aseptic filling operations and limiting the opportunity for post-heat-process contamination by non-proteolytic types of *C. botulinum* is clearly critical. This is often achieved by so-called 'clean-fill' precautions which may include filling under an 'air curtain' using previously steam sterilised production equipment. Whichever method is adopted, procedures to ensure that post-heating contamination does not occur are essential.

Product is formulated to prevent the growth of *C. botulinum*

Many products for which *C. botulinum* is a potential hazard in the raw ingredients are made safe by the presence of some form of preservation agent. This may be by the addition of salt, an organic acid (added artificially or produced naturally by fermentation), drying or some other preservation system.

If the formulation of the product is to be relied upon to achieve safety, it is essential that the factors and their associated levels/concentrations necessary for safety are known and controlled. Where acids or humectants such as salt or sucrose are added to confer safety, then the appropriate amounts necessary to ensure safety must be added and distributed effectively throughout the product. Failure to ensure that correct concentrations are achieved throughout the product mix will compromise the safety of the final product. Methods of manufacture and systems for monitoring the distribution and concentration of controlling factors must be in place.

The choice of the correct acid and appropriate pH is also important and any drift that may occur in the product pH during the life of the product due to interactions with other ingredients or growth of other contaminating microorganisms must be assessed and taken into account.

Where control is achieved by drying, to reduce moisture, it is essential that this is conducted in a way that ensures growth of *C. botulinum* does not occur during the drying stage, which usually involves periods at elevated temperatures. The process should be monitored and the final product water activity should be determined to ensure the correct level is attained.

In many cases it is important to establish the relevant parameter(s) that should be measured to give a reliable indication of product safety and stability. In most cases where water availability is the critical parameter, then the final water activity and not the moisture content is the most relevant factor to measure as an indication of product safety. However, for some products the measurement of water activity is not sufficiently discriminatory or reliable for differentiating between safe and unsafe formulations, as in the case of processed cheese and processed cheese spread, which are usually monitored by measurement of aqueous salt content (and pH) rather than water activity.

In more traditional products such as salamis and dried meats, the factors contributing to safety change during the course of the production process, including the reduction of pH, production of organic acids by lactic acid bacteria, reduction in water activity caused by the addition of salt and reduction of moisture by drying. Although highly traditional processes, it is still important that the key parameters subject to change during the process are clearly established and the process conditions causing these changes are clearly understood and operated correctly. Factors such as chamber temperature and humidity and consequent reduction in pH and weight of the product (as an indication of moisture loss) should be under tight control and regularly monitored to ensure the consistent production of a safe finished product.

Consumer issues and advice to susceptible groups

Control of *C. botulinum* often extends beyond the manufacturer and retailer and into the consumer's home. Many products rely on effective handling of the product by the consumer to ensure they remain safe up to the point of consumption. Chilled products, susceptible to the growth of non-proteolytic *C. botulinum*, must be consumed within the indicated shelf life otherwise unsafe levels of the organism may develop. Likewise, products may be processed and formulated to control non-proteolytic strains of *C. botulinum* in chilled products but, if temperature abused for extended periods, the formulation may not prevent the growth of pro-teolytic strains of the organism, which are tolerant of harsher preservation conditions. Outbreaks of botulism caused by temperature abuse of product in the home have occurred on a number of occasions.

If control measures need to be applied by the consumer, it is essential that full information is presented to consumers about what is required to ensure these are carried out effectively. Such information is best pre-sented in the form of on-pack instructions, clearly detailing the require-ment, for example, for effective temperature control or for consumption of a food within a certain time period. On-pack labelling to 'Keep refri-gerated' and 'Consume within' a defined time period should be displayed prominently on the pack, in large, clear, legible print.

Where control of the hazard is not possible and the risk from such foods to susceptible groups is evident, it is essential to advise consumers of this risk. This occurs very rarely indeed with *C. botulinum* but has occurred in the case of honey. Honey is one of the few foods with an apparent association with infant botulism which occurs following the consumption by young infants of spores of the organism in honey. The organism germinates and grows in the immature infant's intestine with consequent in-situ produc-tion of botulinum toxin, resulting in botulism. Due to this risk, although extremely low, honey manufacturers and retailers now label honey as being 'Unsuitable for infants under the age of 12 months' as a precaution against the hazard presented by this organism in this specific food.

Informing the purchaser and consumer of potential hazards with different foods and their role in controlling such hazards is an essential element of food safety strategies which must encompass all those in the food man-ufacturing, supply, preparation and consumption chain.

5

INDUSTRY ACTION AND REACTION

INTRODUCTION

The approach taken by the food industry to *C. botulinum* differs to that taken in respect of any other foodborne bacterial pathogen in that hazard analysis, process controls and process monitoring are relied upon for product safety, without the use of specific microbiological tests carried out to detect the presence of the organism at any stage of the process. Where microbiological tests are used, these are generally directed at monitoring general spore loads in raw material batches or the presence and level of different microbial groups for monitoring process controls, e.g. coliforms or total Enterobacteriaceae to monitor hygienic practice.

The hazard represented by *C. botulinum* has been recognised and understood by the food industry, particularly the food canning/bottling sector, for many decades and the continued occurrences of outbreaks of foodborne botulism serve as intermittent reminders of its presence. In addition, modified atmosphere packaging (including vacuum packaging) of a wide range of raw and cooked foods in order to maintain the quality of the food during extended chilled shelf life has developed and grown over the past twenty to thirty years to be an important area of food technology and continues to grow, e.g. the development over the last few years of sous-vide products.

In any hazard analysis of such products, *C. botulinum* is a primary target for consideration. Despite this, the food industry has not and still does not include this organism among the pathogens regularly featured in food product specifications such as *Listeria monocytogenes*, *Salmonella* spp. and coagulase positive *staphylococci*. Instead, the likely presence of spores of *C. botulinum* is generally accepted and processes, product formulation, storage temperatures and product shelf life are determined and implemented to destroy or prevent the growth of the organism and subsequent production of toxin.

Heat processes used are based on those already widely accepted, e.g. 12*D* process for shelf-stable, unpreserved canned foods, 6*D* process for spores of psychrotrophic *C. botulinum* in chilled foods (Hersom and Hulland, 1980; Advisory Committee on the Microbiological Safety of Food, 1992). For chilled products, storage temperatures and product shelf lives applied to different products are generally in-line with industry codes of practice, e.g. the UK Chilled Food Association Guidelines for Good Hygienic Practice in the Manufacture of Chilled Foods (Chilled Food Association, 1997).

Safe product shelf life, however, is also linked to product formulation. The presence, level and combinations of certain growth inhibitory substances, e.g. sodium chloride and sodium nitrite, may allow an extended shelf life to be given. In these areas of uncertainty, food industry microbiologists sometimes employ the use of product challenge tests to provide information to demonstrate product safety with respect to *C. botulinum.*

Challenge tests are still used by the canning industry to confirm the efficacy of heat processes applied to both shelf-stable and chilled, long-life canned foods where the product contains preservatives. They are also used by the chilled food manufacturing sector to determine safe product shelf life under specified conditions of product formulation, process and storage temperature/time.

A thorough and well-documented hazard analysis, appropriate challenge test information, implementation and maintenance of critical control point monitors, detailed production process records and appropriate action procedures in the event of process failure all go towards forming the basis of a structured approach to the consistent safety of products for which *C. botulinum* is a potentially significant hazard. Any modification to product ingredients, processes, equipment or manufacturing procedures must be examined for any effect on the hazard of *C. botulinum* and appropriate action taken where these are identified. Failure to do so can lead to disastrous consequences, as in the outbreak of *C. botulinum* associated with hazelnut yogurt (O'Mahony *et al.*, 1990) described earlier.

Food industry and food industry regulatory bodies do not directly apply microbiological criteria containing specific reference to *C. botulinum* or its toxins in standards, guidelines or specifications. Various national and international legislation and industry codes of practice, however, do refer to requirements for implementing specific approaches to processes and control systems targeted at minimising the risk of outbreaks of foodborne botulism.

LEGISLATION AND STANDARDS

There appears to be no national or international legislation specifically referring to criteria for the absence of *C. botulinum* or its toxins from foods. The UK Dairy Products (Hygiene) Regulations (1995) (Anon, 1995a) implementing the European Union Council Directive 92/46/EEC laying down the health rules for the production and placing on the market of raw milk, heat-treated milk and milk-based products does make a general requirement in Schedule 6, part 1, that 'on removal from the processing establishment milk-based products shall not contain pathogenic microorganisms and toxins from pathogenic microorganisms in such quantity as to affect the health of the ultimate consumer.' Other specific food/water legislation contains similar statements and such requirements must naturally include *C. botulinum* and its toxins. In addition, more generic legislation makes it an offence for food to be sold which is not of the nature or substance or quality demanded by the purchaser (Anon, 1990a).

The general approach taken to food safety legislation in Europe and North America is to indicate the clear responsibility of food business proprietors for producing and supplying safe and wholesome foods. For instance, in the UK, The Food Safety (General Food Hygiene) Regulations, 1995 Section 4(1) (Anon, 1995b) which implement parts of the European Union Directive 93/43/EEC of 14th June 1993 on the hygiene of foodstuffs (Anon, 1993) state:

> 'A proprietor of a food business shall ensure that any of the following operations, namely, the preparation, processing, manufacturing, packaging, storing, transportation, distribution, handling and offering for sale or supply, of food are carried out in a hygienic way.'

Further, in Section 4(3):

> 'A proprietor of a food business shall identify any step in the activities of the food business which is critical to ensuring food safety and ensure that adequate safety procedures are identified, implemented, maintained and reviewed on the basis of the following principles:
> (a) analysis of the potential food hazards in a food business operation;
> (b) identification of the points in those operations where food hazards may occur;
> (c) deciding which of the points identified are critical to ensuring food safety ("critical points");
> (d) identification and implementation of effective control and monitoring procedures at those critical points; and

(e) review of the analysis of food hazards, the critical points and the control and monitoring procedures periodically, and whenever the food business's operations change.'

Clearly the severe nature of botulism makes *C. botulinum* an essential consideration in the hazard analysis of many food business operations.

In addition to the general but important and necessary responsibility imposed by legislation on food business proprietors, other legislation (sometimes referred to as vertical legislation because it deals with a specific food in contrast to horizontal legislation which applies to generic food production controls, e.g. food hygiene) may also apply depending on the food type and business. It is obviously the responsibility of the food business proprietor to know and understand which legislation applies to the business and ensure compliance.

The UK Advisory Committee on the Microbiological Safety of Food summarised the relevant UK, EU, and international regulations and codes of practice existing at the time they reported on 'Vacuum packaging and associated processes' (Advisory Committee on the Microbiological Safety of Food, 1992).

In Council Directives of the European Union, hygiene and temperature control in food preparation and distribution and the adoption of HACCP-based systems for identifying hazards and determining, operating and monitoring means for their control are the basis of the approach to the microbiological safety of food.

In France, the Ministry of Agriculture requires registration of premises preparing chilled ready-to-eat meals with shelf lives up to six days. The operators of such premises may be authorised to extend the shelf life of chilled (kept at $3°C$), ready-to-eat meals, heat-treated in the final pack to up to 42 days provided certain conditions relating to heat treatment applied and microbiological criteria are met. The requirement for shelf life temperature to be maintained at $3°C$ is clearly targeted at the control of growth and toxin production by *C. botulinum*.

In New York State, USA, specific restrictions which refer to product characteristics, e.g. food type, pH, water activity and level of competing flora, are in place concerning vacuum-packaged or modified atmosphere packaged foods; also, a licence is required for vacuum or modified atmosphere packaging to be carried out at retail level but only of foods which will not support the growth of *C. botulinum*. Similar restrictions are operated by Maryland State Health Department, USA, and advocated by the US Food and Drug Administration (FDA). In 1983, the FDA also published definitions and good manufacturing practice requirements (FDA Code of Federal Regulations, 21, CFR 113.3) relating to thermally processed low-acid foods packaged in hermetically sealed containers.

Much of the national and international official documentation relating specifically or generally to the control of *C. botulinum* in foods is, however, not in the form of 'Regulations' but as Guidelines or Codes of Practice.

GUIDELINES AND CODES OF PRACTICE

A variety of guidelines have been produced over the past two decades or so dealing with the safe production of canned foods. In the UK in 1981, a Food Hygiene Code of Practice (No. 10) concerning the canning of low-acid foods was issued under the then Food and Drugs Act, 1955. The code has since been replaced by guidelines developed in conjunction with industry dealing with heat preserved foods (Department of Health, 1994). The guidelines advocate a HACCP-based approach to the safe production of heat-preserved foods and include aspects of construction and facilities of manufacturing premises, personnel policy, raw material quality, design and operation of equipment, container types, selection, handling and closure issues, thermal processes, control systems, monitoring systems, product testing and emergency procedures.

Other guidelines have been generated concerning vacuum-packaged or modified atmosphere-packaged, chilled foods. Following a review of factors for use in preventing growth and toxin production by psychrotrophic *C. botulinum*, the UK Advisory Committee on the Microbiological Safety of Food (1992) made recommendations 'that, in addition to chill temperatures which should be maintained throughout the chill chain, the following controlling factors should be used singly or in combination to prevent growth and toxin production by psychrotrophic *C. botulinum* in prepared chilled foods with an assigned shelf life of more than 10 days:

- a heat treatment of 90°C for 10 minutes or equivalent lethality,
- a pH of 5 or less throughout the food and throughout all the components of complex foods,
- a minimum salt level of 3.5% in the aqueous phase throughout the food and throughout all components of complex foods,
- an a_w of 0.97 or less throughout the food and throughout all components of complex foods,

- a combination of heat and preservative factors which can be shown consistently to prevent growth and toxin production by psychrotrophic *C. botulinum.*'

These same parameters and combinations are noted in an illustration of approaches to the control of psychrotrophic *C. botulinum* in food that has not been subject to a heat treatment equivalent to 90°C for 10 minutes and supplied packed in a reduced oxygen atmosphere with a refrigerated shelf life of more than 10 days (Codex Alimentarius Commission, 1997). It is important to ensure that any combination of factors proposed for use are demonstrated to inhibit the growth of psychrotrophic strains of *C. botulinum* within the shelf life and expected storage conditions of the product.

Unless scientific evidence already exists relating to the specific food product type for which *C. botulinum* is considered a hazard, carefully applied predictive mathematical models may be useful or challenge studies could be used to confirm the effectiveness of the selected combined factors against relevant strains of the organism.

The UK Advisory Committee on the Microbiological Safety of Food (1992) and Betts (1996) reviewed national and international codes of practice dealing directly or indirectly with aspects of control of *C. botulinum* in some foods (Table 5.1).

Clearly targeted guidelines or codes of practice written with the direct influence and involvement of the relevant industry manufacturing expertise including production, technology and microbiology personnel are more likely to be of long-term benefit to industry, regulatory authorities and the consumer than non-specific and bureaucratically born legislation. It is therefore in the interests of the food industry to continue to share knowledge and expertise to support the development of industry codes, particularly in important generic areas such as food microbiological safety including the control in foods of *C. botulinum* and its toxins.

SPECIFICATIONS

Product specifications drawn up between a food manufacturer and customer (often a retailer) usually include information concerning the physical appearance of the product, physico-chemical characteristics of importance to the safety and/or quality of the product and microbiological parameters relevant to the safety and quality of the product. However, in products for which *C. botulinum* has been determined a potential hazard, the organism is not included in microbiological specifications for the product but key growth controlling factors are specified and relevant monitoring procedures for these are established and operated.

It is essential, therefore, to ensure that due attention is paid to *C. botulinum* in a thorough hazard analysis carried out and including all aspects of the food production process. Where the organism is considered to represent a potential risk, then, using information from any relevant predictive mathematical modelling or challenge testing normally carried out during the product development stages to help establish product safety parameters, the necessary growth controlling factors must be clearly identified and specified together with required monitoring and action procedures.

MONITORING FOR *CLOSTRIDIUM BOTULINUM*

The food industry does not monitor foods for the specific presence of *C. botulinum* or its spores. Tests to enumerate total aerobic or anaerobic spores are carried out to determine the spore load going in to some specific heat processes, e.g. canned vegetables and unpreserved canned meats, and some tests for the presence of specific spoilage anaerobes may be carried out.

Finished product incubation tests are commonly carried out on canned foods, which may or may not be microbiologically examined after incubation. If cans are assessed only for visual evidence of 'blowing', i.e. swollen cans due to gas production, then non-gas-producing organisms will not be detected. Even so, if incubated can contents are examined microbiologically, only conventional tests will be carried out and no examination made for the specific presence/growth of *C. botulinum* or presence of botulinum toxin. Such specific examinations are not advocated for normal canned foods nor should they be.

There has been a tendency in the recent past (still persisting in some areas of the industry) to establish and maintain a fixed list of microorganisms including potential bacterial pathogens as well as specified non-pathogens and general microbiological tests such as total colony counts and coliforms, which is then applied to all finished product specifications regardless of relevance to the product or the processes by which it was made.

Enumeration tests for *Clostridium perfringens* or sulphite-reducing clostridia are often included in such lists. Some industry personnel have

Table 5.1 Some national and international guidelines and codes of practice relating to aspects of control of *C. botulinum* in foods, adapted from Advisory Committee on the Microbiological Safety of Food (1992) and Betts (1996)

Country/organisation	Status of publication	Foods	Approach includes	Reference
UK Department of Health	Recommended practices	Fresh, hot-smoked and frozen trout	Brining and smoking conditions, storage temperature and shelf life	Department of Health (1978)
UK Department of Health	Guidelines	Cook-chill and cook-freeze catering systems	Hygiene, cooking and cooling conditions, HACCP	Department of Health (1989)
UK Ministry of Agriculture, Fisheries and Food (MAFF)	Advisory note	Smoked fish	Brining and smoking conditions, aqueous salt concentration, chill temperature control	Anon (1991b)
British Meat Manufacturers' Association	Standards	Bacon and bacon joints	Hygienic practices and storage temperatures	Anon (1991c)
Campden and Chorleywood Food Research Association, UK	Code of practice	Vacuum and modified atmosphere packaged chilled foods including sous-vide	Application of HACCP, validation of processes, physico-chemical control parameters, packaging, distribution, storage and shelf life conditions, process auditing	Betts (1996)
UK Chilled Food Association	Guidelines	Chilled foods	Application of HACCP, hygiene standards, quality of ingredients, heat processes to produce a 10^6 reduction of psychrotrophic *C. botulinum* type B, product characteristics inhibitory to the growth of psychrotrophic *C. botulinum* within the shelf life and under the storage conditions of the product	Chilled Food Association (1997)

Table 5.1 Continued

Country/organisation	Status of publication	Foods	Approach includes	Reference
UK Sous Vide Advisory Committee	Code of practice	Sous-vide catering systems	Hygiene standards, quality of ingredients, storage conditions, packaging specifications, heat process and equipment, cooling conditions, shelf life and storage temperature	Anon (1991d)
US National Advisory Committee on Microbiological Criteria for Foods	Recommendations	Refrigerated foods containing cooked, uncured meat or poultry products that are packaged for extended shelf life and that are ready-to-eat or prepared with little or no additional heat treatment	Verified HACCP system to be operated addressing the control of *C. botulinum*	Anon (1990b)
US National Food Processors Association	Guidelines	Refrigerated foods	Application of HACCP, application of challenge testing, safety factors other than refrigeration, shelf life conditions and labelling	National Food Processors Association (1989)
European Chilled Food Federation	Guidelines	Chilled foods	Foods are categorised based on raw/ cooked components, heat process, controlling factors and required consumer handling. Hygienic requirements for 13 categories of foods are discussed in relation to psychrotrophic *C. botulinum*	European Chilled Food Federation (1995)
France	Regulation	Shelf life of ready-to-eat meals	Specifies process and storage temperature requirements for products with extended shelf life up to 21 or 42 days. Compliance with microbiological specifications required	Anon (1988)

considered the latter test to specifically indicate the potential presence of *C. botulinum*. This, of course, is not a sound practice as the methods employed for enumerating sulphite-reducing clostridia 'capture' a wide variety of species of *Clostridium*. Indeed, other sulphite-reducing organisms will also be detected by the method, e.g. some Enterobacteriaceae, especially when a heat treatment stage is not used in the test method. However, such tests may provide useful information about any increases in loading of these types of microorganisms which may, in themselves, be undesirable in the product.

Properly conducted hazard analysis of a food production process supported, if necessary, by information from a well designed challenge test or from the careful application of appropriate predictive mathematical models will determine whether any hazard is posed by *C. botulinum* to the manufactured product. Where a hazard potential is confirmed, it is accepted that the organism will be present from time to time and require routinely applied control measures. The key physico-chemical control factors in the process and of the finished product are therefore determined, specified and routinely monitored. It is also clearly of great importance to ensure that staff are adequately trained to maintain the systems of control, monitoring and to react appropriately to systems failures.

6

TEST METHODS

Efforts have been made to develop conventionally based selective and/or differential tests for detecting *C. botulinum* using, for example, fluorescent antibody-based tests or media incorporating antibiotics with or without specific antitoxins (Hobbs *et al.*, 1982; Silas *et al.*, 1985). Cross reactions with other organisms, lack of sensitivity or otherwise poor discrimination have not been overcome and there are still no routinely used conventional microbiological tests that are specific for detecting *C. botulinum*.

Although it is not difficult to grow *C. botulinum* in a laboratory using conventional methods, media and conditions for growing anaerobic spore-forming bacteria, it is difficult to identify the organism using conventional microbiological tests because it exhibits no biochemical or fermentative properties that distinguish it from some other clostridial species (Advisory Committee on the Microbiological Safety of Food, 1992). Indeed, *C. sporogenes* is phenotypically indistinguishable from Group I *C. botulinum*; the only feature distinguishing it is that it does not produce a neurotoxin.

The original species classification of *C. botulinum* was on the basis of the production of characteristic botulinal neurotoxin even though classification using the more conventional approach of phenotypic characteristics and DNA homology would yield four different groups (Table 1.1). Complications in identification occur because of the existence of species of *Clostridium*, such as *C. sporogenes* (a saprophytic organism commonly found on raw foods), that cannot be distinguished from *C. botulinum* by conventional microbiological tests and, also, species of *Clostridium* other than *C. botulinum* that produce botulinum toxin, e.g. *C. butyricum* and *C. baratii*. However, these two species may be differentiated from each other and *C. botulinum* by, for example, lecithinase production and some sugar reactions (Table 6.1).

Table 6.1 Biochemical characteristics for differentiating botulinum toxin-producing species of *Clostridium*, adapted from Cato *et al.* (1986) and Collins and East (1998)

Characteristic	C. botulinum group				C. baratii	C. butyricum
	I	II	III	IV	F	E
Toxin types	A, B, F	B, E, F	C, D	G		
Lecithinase produced	-	-	±	-	+	-
Liquefaction of gelatin	+	+	+	+	-	-
Fermentation of:						
glucose	+	+	+	-	+	+
fructose	±	+	±	-	+	+
mannose	-	+	+	-	+	+
maltose	±	+	±	-	+	+
sucrose	-	+	-	-	+	+
trehalose	-	+	-	-	-	+
xylose	-	-	-	-	-	+
Lipase	+	+	+	-	-	-
Metabolic acids:						
acetic	+	+	+	+	+	+
propionic			+			
butyric	+	+	+	+	+	+
isobutyric	+			+		
isovaleric	+			+		
phenylacetic				+		

- = negative.

+ = positive.

± = some strains negative, some strains positive.

Some work has been carried out to develop rapid identification systems for *Clostridium* spp. Brett (1998) studied the ability of one biochemical test kit to identify *C. botulinum*. The kit contains a cupule system with substrates and reagents selected for the identification of anaerobes including many species of *Clostridium*. Of 42 strains of *C. botulinum*, four strains of *C. sporogenes* and one strain each of botulinum toxin-producing strains of *C. butyricum* and *C. baratii*, all *C. botulinum* strains were correctly identified to the genus level but species level identification was poor, with 13 strains of *C. botulinum* identified as *C. sporogenes* (>99% discrimination) or *C. histolyticum* (86%) and 17 strains were identified as *C. botulinum* but only at a low level of discrimination, i.e. <50%. All *C. sporogenes* were identified correctly.

C. botulinum type A and proteolytic strains of types B and F have a high DNA–DNA homology with *C. sporogenes* and cannot be distinguished from it either metabolically or biochemically. Toxin neutralisation tests in mice or cell protein electrophoresis patterns can distinguish *C. sporogenes* from the proteolytic botulinum strains but the latter test cannot identify toxin type (Cato *et al.*, 1986).

Because of the significant concern about *C. botulinum* and its toxins in some sectors of the food industry, it is important to have a test or tests available for detecting the toxins of *C. botulinum*. Such tests are particularly valuable for use in the challenge experiments, which may be appropriate during product development to confirm the safety of food compositions.

The acute toxicity test performed in mice is still the only test currently available which provides confident detection of botulinum toxins. It is a very sensitive test measuring biological activity of the toxins but, because it is a non-specific test, to identify any specific toxin type present, antibody neutralisation tests have to be carried out in parallel with the toxicity test. This makes such tests expensive and time-consuming. The tests also require a number of laboratory animals which is a matter of public concern. The US Association of Official Analytical Chemists published a final action, official method for the detection of *C. botulinum* and its toxins in foods based on the mouse toxicity test in 1979 (AOAC, 1995); the method published in the US Food and Drug Administration's Bacteriological Analytical Manual (Kautter *et al.*, 1992) is based on the AOAC method but additional practical information is given for the users of the test. Apart from the culture incubation temperatures indicated, i.e. 26°C and 35°C, the US approach is very similar to the methods used in the UK and elsewhere where a culture incubation temperature of 30°C is more commonly used (Cann *et al.*, 1965; Huss *et al.*, 1974a and b; Nakano *et al.*, 1992; Carlin and Peck, 1995).

Great care and expertise is required to carry out and interpret mouse bioassays as toxicity may be due to components in the product extract other than botulinum toxin. In most cases therefore, it is important to run controls and, as appropriate, to carry out toxin neutralisation studies. Such tests using animals require a special licence and this is only granted where facilities and staff expertise are considered suitable by an authorised inspector. For this reason and the cost of maintaining such facilities, there are relatively few laboratories in the world where animal tests for detecting botulinum toxins may be carried out.

Because of the limitations posed by animal tests, e.g. costs, time, public unease about the use of animals, alternative *in vitro* assays have been developed. These are generally enzyme linked immunosorbent assay (ELISA) methods using monoclonal antibodies to capture the toxin followed by some form of signal amplification (Shone *et al.*, 1985; Potter *et al.*, 1993; Doellgast *et al.*, 1993 and 1994). These tests are not generally as sensitive as the mouse toxicity test and give no measure of biological activity of the toxin. Although these tests have been used experimentally (Carlin and Peck, 1995), they do not appear to be widely used or accepted as a replacement for the mouse toxicity test.

In the past ten years, significant progress has been made in understanding the specific mode of action of botulinum neurotoxins, all of which have been shown to have zinc-dependent protease activity by which each toxin acts highly specifically to cleave a single protein at a single site (Schiavo *et al.*, 1992a, 1992b, 1994; Binz *et al.*, 1994; Yamasaki *et al.*, 1994). This new knowledge is being exploited in the development of specific and sensitive immunoassays for detecting botulinum toxins (Hallis *et al.*, 1996). Indeed, work funded by the UK Ministry of Agriculture, Fisheries and Food and being carried out at the Centre for Applied Microbiology & Research in the UK has successfully developed kits for detecting a range of botulinum toxins; inter-laboratory trials using the kits are currently in progress (Wictome *et al.*, 1999).

Aranda *et al.* (1997) noted that in the last decade, gene sequences for several botulinum neurotoxins have been determined. Using this information, polymerase chain reaction based methods have been used successfully to detect neurotoxin genes (Campbell *et al.*, 1993; Szabo *et al.*, 1993 and 1994; Aranda *et al.*, 1997). Although such methods may be useful, they are likely to be of limited value because the live organism must be present from which the genes encoding botulinum neurotoxins are extracted; the toxins are not specifically detected.

The further development of these newer technologies based on the steadily increasing detailed knowledge and understanding of the genetic structure of all neurotoxin-producing clostridia and modes of action of the neurotoxins will undoubtedly allow *in vitro* methods to be developed that will have equal or better sensitivity than the mouse assay. They may even provide some indication of the potential for their biological activity, which is important where food product safety information is required relating to identified potential hazards from *C. botulinum* (Wictome *et al.*, 1999).

While, for the time being, the investigation of suspected botulism outbreaks will need to be carried out using the sensitive and specific mouse bioassay, it may be possible in the near future to use *in vitro* methods both for this purpose and for food product challenge test studies.

7

THE FUTURE

From the botulism outbreak and food and environment survey information already available, some of which are described in this book, it is already clear that the primary production environments of many food industry raw materials, particularly animal and vegetable/salad vegetable, contribute greatly to a widespread low incidence of *C. botulinum* in these raw materials.

Over the past decade, commercial considerations have driven food product manufacturers and retailers to become increasingly innovative in their product development programmes. The ready availability of food raw materials from anywhere in the world leading to the use of more 'exotic' ingredients and the closure of the 'season gap' which has made normally seasonal foods available all year round have given free rein to this innovative programme.

The market for chilled food products has grown impressively over the last decade, including savoury ready meals (recipe dishes), dairy and dessert products, sandwiches and other snack meals. As evidenced by the outbreak of botulism in Italy attributed to commercially produced mascarpone cheese (Aureli *et al.*, 1996), such products require just as much consideration in a well structured and thorough hazard analysis as those more traditionally regarded as potential botulinum hazards, e.g. shelf-stable canned vegetables, fish or meat.

Increasingly novel and complex combinations of raw materials (meats, poultry, game, fish and shellfish, milk and milk products, eggs, vegetables, salads, grains, nuts, herbs and spices) are being used to produce ranges of consumer-tempting products in large-scale commercial systems. New or alternative food processing technologies, packaging technologies and storage and distribution systems are also being developed and new applications of existing technologies are being explored for use in food production processes. Ohmic heating processes, irradiation, ultra-filtration, high pressure and high intensity light are examples of some newer processing technologies.

These all create new challenges to food technologists and microbiologists who must ensure that due consideration is given to *C. botulinum* during the hazard analysis process that must be applied to each new product and any product where changes are made to ingredients, process or physico-chemical conditions affecting product safety. Where indicated by the hazard analysis or by use of predictive mathematical modelling, appropriately devised and relevant challenge tests should be carried out using the real food.

Even though food-associated botulism is a comparatively rare illness, because of the growth of novel food products, technologies and packaging, there is a clear need for the food industry (manufacturers and retailers) to develop and support a 'routine' rather than *ad hoc* approach to *C. botulinum* hazard assessment in which challenge tests and predictive mathematical modelling are more frequently and appropriately used.

There is now clear evidence that other species of *Clostridium* produce botulinum toxin (McCroskey *et al.*, 1986 and 1991). The long-standing acceptance of the species *C. botulinum* as encompassing any organism that produces a botulinum neurotoxin despite considerable phenotypic differences between groups of strains within the species (Tables 1.1 and 6.1) should now be properly challenged. International co-operation is needed to re-examine and re-define the nomenclature of all the organisms currently grouped within *C. botulinum* and those other organisms found to produce botulinum toxins. In addition to phenotypic data and botulinum neurotoxin production, the increasing genotypic data becoming available should be taken into account in this necessary re-examination of the classification of *C. botulinum* and its 'relatives' (Collins and East, 1998).

In any event, for food microbiologists, it will always be essential that a detailed and competent hazard analysis is carried out at an early stage in all new food product and process developments. This will help to ensure that relevant critical controls and monitoring systems are put in place to minimise potential public health problems which could arise from the presence and outgrowth of *C. botulinum*.

GLOSSARY OF TERMS

Biotyping The conventional method for distinguishing between bacterial types using their metabolic and/or physiological properties (biotype).

Commensal Animals or plants which live as tenants of others and share their food.

D value The time required (usually expressed in minutes) at a given temperature to reduce the number of viable cells or spores of a given microorganism to 10% of the initial population. When used in relation to irradiation, it is the dose in kGy required at a given temperature to reduce the number of viable cells or spores of a given microorganism to 10% of the initial population.

F_0 The time in minutes required at 121.1°C to destroy a specified population of microbial spores or cells. The $F_0 3$ process used in canning is a heat process of 121.1°C for 3 minutes or an equivalent heat process specified to achieve a >12 log cycle reduction in spores of *C. botulinum*.

Genotyping Methods used to differentiate bacteria based on the composition of their nucleic acids.

Guidelines See 'Microbiological guidelines'.

Hazard A biological, chemical or physical agent in, or condition of, food with the potential to cause an adverse health effect (Codex Alimentarius Commission, 1996).

Humectant A substance which absorbs moisture making water less available for microbial growth.

Incertae sedis Of uncertain taxonomic position.

Indicator organism Those organisms whose presence suggests inadequate processing/control for safety.

kGy (kilo gray) = 1000 gray. The gray (Gy) is the SI unit of absorbed dose of ionising radiation = 1 joule of energy absorbed per kilogram of matter. 1 Gy = 100 rad; 10 kGy = 1 megarad (the rad is the traditional unit of absorbed dose of ionising radiation = 100 ergs of energy absorbed per gram of matter).

Lethal Dose$_{50}$ (LD$_{50}$)   The dose of toxin which, when administered to a number of test animals, kills 50% of those animals under the test conditions. MLD$_{50}$ = mouse lethal dose$_{50}$.

Microbiological guidelines   These are criteria applied at any stage of the food production and distribution system to indicate the microbiological condition of a sample. They are for management information and to assist in the identification of potential problem areas.

Microbiological specifications   These are microbiological criteria applied to individual raw materials, ingredients or the end product. They are used in purchase agreements.

Microbiological standards   These are microbiological criteria contained in a law. Compliance is mandatory. Examples include most criteria in European Union (EU) Directives and Statutory Instruments of England and Wales. Standards are monitored by enforcement agencies.

Modified Atmosphere Packaging (MAP)   The intentional modification of the atmosphere surrounding the food. This is carried out at the point of packaging the food usually in pre-formed, gas-impermeable, plastic trays from which air is removed and a gaseous mixture, usually of carbon dioxide and nitrogen, added. At the same time, the pack is covered with a layer of gas-impermeable plastic sheet and heat-sealed to the base tray.

Pasteurisation   A form of heat treatment that kills vegetative pathogens and spoilage microorganisms in milk and other foods, e.g. for milk a common pasteurisation process is 71.7°C for 15 seconds.

Pathogen   Any microorganism which by direct interaction with (infection of) another organism causes disease in that organism.

Phenotype   The observable characteristics of an organism which include biotype, serotype, phage type and bacteriocin type.

Polymerase Chain Reaction (PCR)   A technique used to amplify the number of copies of a pre-selected region of nucleic acid to a sufficient level for detection and testing.

Pulsed Field Gel Electrophoresis (PFGE)   A technique which allows chromosomal restriction fragment patterns to be produced and used as a means of sub-dividing microbial species.

Redox potential (oxidation-reduction potential) A measure of the tendency of a given system to donate electrons (act as a reducing agent) or accept electrons (act as an oxidising agent). The redox potential (Eh) of a particular system may be determined by measuring the electrical potential difference between the system and a standard hydrogen electrode. Eh is often recorded in millivolts (mV). Conditions of temperature and pH at the point of measurement are important and should also be recorded (Singleton and Sainsbury, 1987).

Ribotyping A method for characterising bacterial isolates according to their ribosomal RNA pattern (ribotype) and identifying the isolate by comparing the pattern obtained with a database of patterns.

Risk An estimate of the likelihood of a hazard occurring and its potential adverse health effects.

Saprophytic Organism that obtains nutrients from non-living organic matter by absorbing soluble organic compounds.

Sous-vide Usually composite foods pasteurised in a vacuum pack usually intended for catering outlets. Such products are often given an extended shelf life at refrigeration ($<$ or $= 2°C$) temperatures.

Specifications See 'Microbiological specifications'.

Standards See 'Microbiological standards'.

Strain An isolate or group of isolates that can be distinguished from other isolates of the same genus and species by either phenotypic and/or genotypic characteristics.

Taxonomy The grouping and naming of organisms according to their natural similarities or relationships.

Water activity (a_w) A measure of the availability of water for the growth and metabolism of microorganisms. It is expressed as a ratio of the water vapour pressure of a food or solution to that of pure water at the same temperature.

z value The number of Centigrade degrees required for the thermal destruction curve to traverse one log cycle, e.g. number of Centigrade degrees increase ($C°$) required to decrease the D value by 10-fold.

REFERENCES

Adams, M. R., Hartley, A. D. and Cox, L. J. (1989) Factors affecting the efficacy of washing procedures used in the production of prepared salads. *Food Microbiology*, **6**, 69–77.

Advisory Committee on the Microbiological Safety of Food (1992) *Report on Vacuum Packaging and Associated Processes*. HMSO, London, UK.

American Meat Institute (1989) *Good Manufacturing Practices – Fermented Dry and Semi-Dry Sausage*. American Meat Institute, Washington, DC, USA.

Angulo, F. J., Getz, J., Taylor, J. P. *et al*. (1998) A large outbreak of botulism: the hazardous baked potato. *The Journal of Infectious Diseases*, **178** (July), 172–177.

Anon (1981) Alert: botulism associated with commercially produced, dried, salted whitefish. *California Morbidity*, 6 November. California Department of Health Services, USA.

Anon (1983) Botulism and commercial pot pie – California. *Morbidity and Mortality Weekly Report*, **32**(3), 39–40, 45.

Anon (1988) *Prolongation of life span of pre-cooked food, modification of procedures enabling authorisation to be obtained*. Veterinary Service of Food Hygiene, Ministry of Agriculture, Paris, France.

Anon (1989) *Garlic products in oil assignment (FY 89)*. Memorandum of the Department of Health and Human Services, USA.

Anon (1990a) *Food Safety Act 1990*. Chapter 16, HMSO, London, UK.

Anon (1990b) *Recommendations for Refrigerated Foods containing Cooked, Uncured Meat or Poultry Products that are Packaged for Extended Shelf life and that are Ready-to-Eat or Prepared with little or no Additional Heat Treatment*. US National Advisory Committee on Microbiological Criteria for Foods, USA.

Anon (1991a) *Food and Drink – Good Manufacturing Practice: A Guide to its Responsible Management*, 3rd edn. Institute of Food Science and Technology, London, UK. (Now in 4th edn, 1998).

Anon (1991b) *Microbiological Safety of Smoked Fish*. Ministry of Agriculture, Fisheries and Food, London, UK.

Anon (1991c) *Standards: 1. For the Production of Bacon and Bacon Joints 2. Accredited Standards of Good Manufacturing Practice*. British Meat Manufacturers' Association, London, UK.

Anon (1991d) *Code of Practice for Sous Vide Catering Systems*. Sous-Vide Advisory Committee, Tetbury, UK.

Anon (1992) *Code for the production of microbiologically safe and stable emulsified and non-emulsified sauces containing acetic acid*. Comité des

Industries des Mayonnaises et Sauces Condimentaires de la Communaute Economique Européenne (CIMSCEE), Bruxelles, Belgium.

Anon (1993) Council Directive 93/43/EEC of 14th June 1993 on the hygiene of foodstuffs. *Official Journal of the European Communities*, 19.7.93, No. L175, 1-11.

Anon (1995a) *The Dairy Products (Hygiene) Regulations, 1995. Statutory Instrument No. 1086.* HMSO, London, UK.

Anon (1995b) *The Food Safety (General Food Hygiene) Regulations 1995. Statutory Instrument No. 1763.* HMSO, London, UK.

Anon (1998a) Organic vegetable soup associated with a case of botulism - northern Italy. *Eurosurveillance Weekly* (10 Sept).

Anon (1998b) Botulism associated with home-preserved mushrooms. *Communicable Disease Report Weekly*, **8**(18), 159, 162.

Anon (1999) Black olives spark botulism alert. *World Food Law* (15 January).

AOAC (1995) *AOAC Official Methods of Analysis, 16th edition, Chapter 17, Subchapter 7, Method 977.26,* Clostridium botulinum *and its toxins in foods.* Association of Official Analytical Chemists, Arlington, Virginia, USA.

Aramouni, F. M., Kone, K. K., Craig, J. A. *et al.* (1994) Growth of *Clostridium sporogenes* PA 3679 in home-style canned quick breads. *Journal of Food Protection*, **57**(10), 882-886.

Aranda, E., Rodriguez, M. M., Asensio, M. A. *et al.* (1997) Detection of *Clostridium botulinum* types A, B, E and F in foods by PCR and DNA probe. *Letters in Applied Microbiology*, **25**, 186-190.

Arnon, S. S. (1980) Infant botulism. *Annual Reviews of Medicine*, **31**, 541-560.

Aureli, P., Fenicia, L., Pasolini, B. *et al.* (1986) Two cases of type E infant botulism caused by neurotoxigenic *C. butyricum* in Italy. *Journal of Infectious Diseases*, **154**, 207-211.

Aureli, P., Franciosa, G. and Pourshaban, M. (1996) Foodborne botulism in Italy. *The Lancet*, **348** (7 Dec), 1594.

Austin, J. W., Dodds, K. L., Blanchfield, B. *et al.* (1998) Growth and toxin production by *Clostridium botulinum* on inoculated fresh-cut packaged vegetables. *Journal of Food Protection*, **61**(3), 324-328.

Baird-Parker, A. C. and Freame, B. (1967) Combined effect of water activity, pH and temperature on the growth of *Clostridium botulinum* from spore and vegetative cell inocula. *Journal of Applied Bacteriology*, **30**(3), 420-429.

Ball, A. P., Hopkinson, R. B., Farrell, I. D. *et al.* (1979) Human botulism caused by *Clostridium botulinum* type E: the Birmingham outbreak. *Quarterly Journal of Medicine*, New Series XLVIII, **191**, 473-491.

Bell, E., Bennett, P., Friedman, S. *et al.* (1985) Botulism associated with commercially distributed Kapchunka - New York City. *Morbidity and Mortality Weekly Report*, **34**(35) (6 Sept), 546-547.

Betts, G. D. and Gaze, J. E. (1992) *Food Pasteurisation Treatments: Part 2 - Recommendations for the Design of Pasteurisation Processes. Technical Manual No. 27.* Campden and Chorleywood Food Research Association, Chipping Campden, UK.

Betts, G. D. and Gaze, J. E. (1995) Growth and heat resistance of psychrotrophic *Clostridium botulinum* in relation to 'sous-vide' products. *Food Control*, **6**(1), 57-63.

Betts, G. D. (ed.) (1996) *Code of Practice for the Manufacture of Vacuum and Modified Atmosphere Packaged Chilled Foods with Particular Regard to the Risks of Botulism, Guideline No. 11.* Campden and Chorleywood Food Research Association, Chipping Campden, UK.

Beuchat, L. R. (1992) Surface disinfection of raw produce. *Dairy, Food and Environmental Sanitation*, **12**(1), 6-9.

Binz, T., Blasi, J., Yamasaki, S. *et al.* (1994) Proteolysis of SNAP-25 by types E and A botulinal neurotoxins. *Journal of Biological Chemistry*, **269**(3), 1617-1620.

Blatherwick, F. J., Peck, S. H., Morgan, G. B. *et al.* (1985a) Update: international outbreak of restaurant-associated botulism: Vancouver, British Columbia, Canada. *Morbidity and Mortality Weekly Report*, **34**(41), 643.

Blatherwick, F. J., Peck, S. H., Morgan, G. B. *et al.* (1985b) An international outbreak of botulism associated with a restaurant in Vancouver, British Columbia. *Canada Diseases Weekly Report*, **11–42**, 177-178.

Bow, R., Ferguson, H., Madson, C. *et al.* (1974) Botulism - Idaho, Utah. *Morbidity and Mortality Weekly Report*, (6 July), 241-242.

Brent, J., Gomez, H., Judson, F. *et al.* (1995) Botulism from potato salad. *Dairy, Food and Environmental Sanitation*, **15**(7), 420-422.

Brett, M. M. (1998) Evaluation of the use of the bioMérieux Rapid ID32 A for the identification of *C. botulinum*. *Letters in Applied Microbiology*, **26**, 81-84.

Brett, M. (1999) Botulism in the United Kingdom. *Eurosurveillance*, **4**(1), 9-11.

Briozzo, J., de Lagarde, E. A., Chirife, J. *et al.* (1983) *Clostridium botulinum* type A growth and toxin production in media and process cheese spread. *Applied and Environmental Microbiology*, **45**(3), 1150-1152.

Brown, G. D. and Gaze, J. E. (1990) *Determination of the Growth Potential of* Clostridium botulinum *types E and Non-proteolytic B in Sous-Vide Products at Low Temperatures. Technical Memorandum No. 593.* Campden Food and Drink Research Association, Chipping Campden, UK.

Brown, G. D., Gaze, J. E. and Gaskell, D. E. (1991) *Growth of* Clostridium botulinum *Non-proteolytic type B and type E in 'Sous-Vide' Products Stored at 2-15°C. Technical Memorandum No. 635.* Campden Food and Drink Research Association, Chipping Campden, UK.

Campbell, K. D., Collins, M. D. and East, A. K. (1993) Gene probes for identification of the botulinal neurotoxin gene and specific identification of neurotoxin types B, E, and F. *Journal of Clinical Microbiology*, **31**(9), 2255-2262.

Cann, D.C. and Taylor, L. Y. (1979) The control of the botulism hazard in hot-smoked trout and mackerel. *Journal of Food Technology*, **14**, 123-129.

Cann, D. C., Wilson, B. B., Hobbs, G. *et al.* (1965) The incidence of *Clostridium botulinum* type E in fish and bottom deposits in the North Sea and off the coast of Scandinavia. *Journal of Applied Bacteriology*, **28**(3), 426-430.

Cann, D. C., Wilson, B. B. and Hobbs, G. (1968) Incidence of *Clostridium botulinum* in bottom deposits in British coastal waters. *Journal of Applied Bacteriology*, **31**, 511-514.

Cann, D. C., Taylor, L. Y. and Hobbs, G. (1975) The incidence of *Clostridium botulinum* in farmed trout raised in Great Britain. *Journal of Applied Bacteriology*, **39**, 331-336.

Carlin, F. and Peck, M. W. (1995) Growth and toxin production by non-proteolytic and proteolytic *Clostridium botulinum* in cooked vegetables. *Letters in Applied Microbiology*, **20**, 152-156.

Carlin, F. and Peck, M. W. (1996) Growth of and toxin production by non-proteolytic *Clostridium botulinum* in cooked puréed vegetables at refrigeration temperatures. *Applied and Environmental Microbiology*, **62**(8), 3069-3072.

Cato, E. P., George, W. L. and Finegold, S.M. (1986) Genus *Clostridium* Prazmowski 1880, 23[AL], in *Bergey's Manual of Systematic Bacteriology*, 9th edn, vol. 2 (eds P. H. A. Sneath, N. S. Mair, M. E. Sharpe and J. G. Holt). Williams and Wilkins, Baltimore, USA, pp. 1141-1160.

CCFRA (1975) *Canning Retorts and their Operation. Technical Manual No. 2.* Campden and Chorleywood Food Research Association, Chipping Campden, UK.

CCFRA (1977) *Guidelines for the Establishment of Scheduled Heat Processes for Low Acid Foods. Technical Manual No. 3.* Campden and Chorleywood Food Research Association, Chipping Campden, UK.

CCFRA (1997a) *Guidelines for Batch Retort Systems - Full Water Immersion - Raining Water - Steam Air. Guideline No. 13.* Campden and Chorleywood Food Research Association, Chipping Campden, UK.

CCFRA (1997b) *Guidelines for Performing Heat Penetration Trials for Establishing Thermal Processes in Batch Retort Systems. Guideline No.16.* Campden and Chorleywood Food Research Association, Chipping Campden, UK.

CCFRA (1997c) *Guidelines for Establishing Heat Distribution in Batch Overpressure Retort Systems. Guideline No.17.* Campden and Chorleywood Food Research Association, Chipping Campden, UK.

Chilled Food Association (1997) *Guidelines for Good Hygienic Practice in the Manufacture of Chilled Foods*, 3rd edn. Chilled Food Association, London, UK.

Chou, J. H., Hwang, P. H. and Malison, M. D. (1988) An outbreak of type A foodborne botulism in Taiwan due to commercially preserved peanuts. *International Journal of Epidemiology*, **17**(4), 899-902.

Christiansen, L. N., Tompkin, R. B., Shaparis, A. B. *et al.* (1975) Effect of sodium nitrite and nitrate on *Clostridium botulinum* growth and toxin production in a summer style sausage. *Journal of Food Science*, **40**, 488-490.

Codex Alimentarius Commission (1996) *Draft Hazard Analysis and Critical Control Point (HACCP) System and Guidelines for its Application.* Appendix II. Alinorm 97/13A. Joint FAO/WHO Food Standards Programme Report of the twenty-ninth session of the Codex Committee on Food Hygiene, 21-25 October 1996, Washington, DC, USA.

Codex Alimentarius Commission (1997) *Proposed Draft Code of Hygienic Practice for Refrigerated Packaged Foods with Extended Shelf Life.* Thirtieth Session of the Codex Commission on Food Hygiene, Washington, DC, USA, 20-24 October 1997, Joint FAO/WHO Food Standards Programme. Twenty-third session of the Codex Alimentarius Commission, 1999, Rome, Italy.

Collins, M. D. and East, A. K. (1998) A review: phylogeny and taxonomy of the food-borne pathogen *Clostridium botulinum* and its neurotoxins. *Journal of Applied Microbiology*, **84**, 5-17.

Collins-Thompson, D. L. and Wood, D. S. (1993) Control in dairy products, in Clostridium botulinum - *Ecology and Control in Foods* (eds A. H. W. Hauschild and K. L. Dodds). Marcel Dekker, Inc, New York, USA, pp. 261-277.

Critchley, E. M. R., Hayes, P. J. and Isaacs, P. E. T. (1989) Outbreak of botulism in north-west England and Wales, June 1989. *The Lancet*, (7 October), 849-853.

Daifas, D. P., Smith, J. P., Blanchfield, B. *et al.* (1999) Growth and toxin production by *Clostridium botulinum* in English-style crumpets packaged under modified atmospheres. *Journal of Food Protection*, **62**(4), 349-355.

D'Argenio, P., Palumbo, F., Ortolani, R. *et al.* (1995) Type B botulism associated with roasted eggplant in oil - Italy, 1993. *Morbidity and Mortality Weekly Report*, **44**(2), 33-36.

del Torre, M., Stecchini, M. L. and Peck, M. W. (1998) Investigation of the ability of proteolytic *Clostridium botulinum* to multiply and produce toxin in fresh Italian pasta. *Journal of Food Protection*, **61**(8), 988-993.

Denny, C. B., Goeke, D. J. and Sternberg, R. (1969) *Inoculation Tests of* Clostridium botulinum *in Canned Breads with Special Reference to Water Activity. Research Report No. 4-69.* Washington Research Laboratory, National Canners Association, Washington, DC, USA.

Department of Health (1978) *Recommended Practices for the Processing, Handling and Cooking of Fresh, Hot-Smoked and Frozen Trout.* HMSO, London, UK.

Department of Health (1989) *Guidelines on Cook-Chill and Cook-Freeze Catering Systems.* HMSO, London, UK.

Department of Health (1994) *Guidelines for the Safe Production of Heat Preserved Foods.* HMSO, London, UK.

Dodds, K. L. (1989) Combined effect of water activity and pH on inhibition of toxin production by *Clostridium botulinum* in cooked, vacuum-packed potatoes. *Applied and Environmental Microbiology,* **55**(3), 656–660.

Dodds, K. L. (1990) Restaurant-associated botulism outbreaks in North America. *Food Control* (July), 139–141.

Dodds, K. L. (1993a) *Clostridium botulinum* in the environment, in Clostridium botulinum – *Ecology and Control in Foods* (eds A. H. W. Hauschild and K. L. Dodds). Marcel Dekker, Inc, New York, USA, pp. 21–51.

Dodds, K. L. (1993b) *Clostridium botulinum* in foods, in Clostridium botulinum – *Ecology and Control in Foods* (eds A. H. W. Hauschild and K. L. Dodds). Marcel Dekker, Inc, New York, USA, pp. 53–68.

Doellgast, G. J., Triscott, M. X., Beard, G. A. *et al.* (1993) Sensitive enzyme-linked immunosorbent assay for detection of *Clostridium botulinum* neurotoxins A, B and E using signal amplification via enzyme-linked coagulation assay. *Journal of Clinical Microbiology,* **31**(9), 2402–2409.

Doellgast, G. J., Beard, G. A., Bottoms, J. D. *et al.* (1994) Enzyme-linked immunosorbent assay and enzyme-linked coagulation assay for detection of *Clostridium botulinum* neurotoxins A, B and E and solution-phase complexes with dual-label antibodies. *Journal of Clinical Microbiology,* **32**(1), 105–111.

Eklund, M. W. (1982) Significance of *Clostridium botulinum* in fishery products preserved short of sterilization. *Food Technology,* December, 107–112, 115.

Esty, J. R. and Meyer, K. F. (1922) The heat resistance of the spores of *B. botulinus* and allied anaerobes. XI. *Journal of Infectious Diseases,* **31**, 650–663.

European Chilled Food Federation (1995) *Guidelines for the Hygienic Manufacture of Chilled Foods.* c/o Chilled Food Association, London, UK.

Fairbrother, R. W. (1938) *A Text-book of Medical Bacteriology,* 2nd edn. William Heinemann (Medical Books) Ltd, London, UK.

Fernandez, P. S. and Peck, M. W. (1999) Predictive model that describes the effect of prolonged heating at 70°C-90°C and subsequent incubation at refrigeration temperatures on growth and toxigenesis by non-proteolytic *Clostridium botulinum* in the presence of lysozyme. *Applied and Environmental Microbiology,* **65**(8), 3449–3457.

Food and Drug Administration (1996) FDA warns against consuming certain Italian mascarpone cream cheese because of potential serious botulism risk. *HHS News* (9 September). US Department of Health and Human Services, Food and Drug Administration, Washington, DC, USA.

Food and Drug Administration (1999) *Food Code, Chapter 3: Food, Section 3-501.16 Potentially Hazardous Food, Hot and Cold Holding.* US Department of Health and Human Services, Public Health Service, Food and Drug Administration, Washington, DC, USA.

Food MicroModel (1999) Version 3.02, Food MicroModel Ltd, Randalls Road, Leatherhead, Surrey, UK. (Models originally funded by the Ministry of Agriculture, Fisheries and Food, UK.)

Franciosa, G., Pourshaban, M., Gianfranceschi, M. *et al.* (1999) *Clostridium botulinum* spores and toxin in mascarpone cheese and other milk products. *Journal of Food Protection*, 62(8), 867-871.

Garren, D. M., Harrison, M. A. and Huang, Y-W. (1995) Growth and production of toxin of *Clostridium botulinum* type E in rainbow trout under various storage conditions. *Journal of Food Protection*, 58(8), 863-866.

Gaze, J. E. (ed.) (1992) *Food Pasteurisation Treatments: Part 1 - Guidelines to the Types of Food Products Stabilised by Pasteurisation Treatments. Technical Manual No. 27.* Campden and Chorleywood Food Research Association, Chipping Campden, UK.

Gaze, J. E., Shaw, R. and Archer, J. (1998) *Identification and Prevention of Hazards Associated with Slow Cooling of Hams and Other Large Cooked Meats and Meat Products. Review No. 8.* Campden and Chorleywood Food Research Association, Chipping Campden, UK.

Gibbs, P. A., Davies, A. R. and Fletcher, R. S. (1994) Incidence and growth of psychrotrophic *Clostridium botulinum* in foods. *Food Control*, 5(1), 5-7.

Glass, K. A. and Doyle, M. P. (1991) Relationship between water activity of fresh pasta and toxin production by proteolytic *Clostridium botulinum*. *Journal of Food Protection*, 54(3), 162-165.

Glass, K. A., Kaufman, K. M., Smith, A. L. *et al.* (1999) Toxin production by *Clostridium botulinum* in pasteurised milk treated with carbon dioxide. *Journal of Food Protection*, 62(8), 872-876.

Graham, A. F., Mason, D. R. and Peck, M. W. (1996a) Predictive model of the effect of temperature, pH and sodium chloride on growth from spores of non-proteolytic *Clostridium botulinum*. *International Journal of Food Microbiology*, 31, 69-85.

Graham, A. F., Mason, D. R. and Peck, M. W. (1996b) Inhibitory effect of combinations of heat treatment, pH, and sodium chloride on growth from spores of non-proteolytic *Clostridium botulinum* at refrigeration temperature. *Applied and Environmental Microbiology*, 62(7), 2664-2668.

Graham, A. F., Mason, D. R., Maxwell, F. J. *et al.* (1997) Effect of pH and NaCl on growth from spores of non-proteolytic *Clostridium botulinum* at chill temperatures. *Letters in Applied Microbiology*, 24, 95-100.

Hallis, B., James, B. A. F. and Shone, C. C. (1996) Development of novel assays for botulinum type A and B neurotoxins based on their endopeptidase activities. *Journal of Clinical Microbiology*, 34(8), 1934-1938.

Hao, Y. -Y., Brackett, R. E., Beuchat, L. R. *et al.* (1998) Microbiological quality and the inability of proteolytic *Clostridium botulinum* to produce toxin in film-packaged fresh-cut cabbage and lettuce. *Journal of Food Protection*, 61(9), 1148-1153.

Hao, Y. -Y., Brackett, R.E., Beuchat, L.R. *et al.* (1999) Microbiological quality and production of botulinal toxin in film-packaged broccoli, carrots and green beans. *Journal of Food Protection*, 62(5), 499-508.

Hauschild, A. H. W. (1989) *Clostridium botulinum*, in *Foodborne Bacterial Pathogens* (ed. M. P. Doyle). Marcel Dekker, Inc, New York, USA, pp. 111-189.

Hauschild, A. H. W. (1993) Epidemiology of human foodborne botulism, in Clostridium botulinum - *Ecology and Control in Foods* (eds A. H. W. Hauschild and K. L. Dodds). Marcel Dekker, Inc, New York, USA, pp. 69-104.

Hauschild, A. H. W. and Simonsen, B. (1985) Safety of shelf-stable, canned, cured meats. *Journal of Food Protection*, **48**(11), 997-1009.

Heinitz, M. L. and Johnson, J. M. (1998) The incidence of *Listeria* spp., *Salmonella* spp., and *Clostridium botulinum* in smoked fish and shellfish. *Journal of Food Protection*, **61**(3), 318-323.

Hersom, A. C. and Hulland, E. D. (1980) *Canned Foods - Thermal Processing and Microbiology*, 7th edn. Churchill Livingstone, London, UK.

Hielm, S., Björkroth, J., Hyytiä, E. *et al.* (1998) Prevalence of *Clostridium botulinum* in Finnish trout farms: pulsed-field gel electrophoresis typing reveals extensive genetic diversity among type E isolates. *Applied and Environmental Microbiology*, **64**(11), 4161-4167.

Hobbs, G., Crowther, J. S., Neaves, P. *et al.* (1982) Detection and isolation of *Clostridium botulinum*, in *Isolation and Identification Methods for Food Poisoning Organisms* (eds J. E. L. Corry, D. Roberts and F. A. Skinner). The Society for Applied Bacteriology Technical Series, No. 17. Academic Press, London, UK, pp. 151-164.

Houtsma, P. C., Heuvelink, A., Dufrenne, J. *et al.* (1994) Effect of sodium lactate on toxin production, spore germination and heat resistance of proteolytic *Clostridium botulinum* strains. *Journal of Food Protection*, **57**(4), 327-330.

Huss, H. H. (1980) Distribution of *Clostridium botulinum*. *Applied and Environmental Microbiology*, **39**(4), 764-769.

Huss, H. H., Pedersen, A. and Cann, D. C. (1974a) The incidence of *Clostridium botulinum* in Danish trout farms. I. Distribution in fish and their environment. *Journal of Food Technology*, **9**, 445-450.

Huss, H. H., Pedersen, A. and Cann, D. C. (1974b) The incidence of *Clostridium botulinum* in Danish trout farms. II. Measures to reduce contamination of the fish. *Journal of Food Technology*, **9**, 451-458.

Hutton, M. T., Koskinen, M. A. and Hanlin, J. H. (1991) Interacting effects of pH and NaCl on heat resistance of bacterial spores. *Journal of Food Science*, **56**(3), 821-822.

Hyytiä, E., Eerola, S., Hielm, S. *et al.* (1997) Sodium nitrite and potassium nitrate in control of non-proteolytic *Clostridium botulinum* outgrowth and toxigenesis in vacuum-packed cold-smoked rainbow trout. *International Journal of Food Microbiology*, **37**, 63-72.

Hyytiä, E., Hielm, S. and Korkeala, H. (1998) Prevalence of *Clostridium botulinum* type E in Finnish fish and fishery products. *Epidemiology and Infection*, **120**(3), 245-250.

Incze, K. (1992) Raw fermented and dried meat products. *Fleischwirtsch. International*, **2**, 3-12.

International Commission on Microbiological Specifications for Foods (ICMSF) (1980) *Microbial Ecology of Foods, Volume 1: Factors Affecting Life and Death of Microorganisms*. Academic Press, London, UK.

International Commission on Microbiological Specifications for Foods (ICMSF) (1996) *Microorganisms in Foods. 5. Microbiological Specifications of Food Pathogens*. Blackie Academic and Professional, London, UK.

International Commission on Microbiological Specifications for Foods (ICMSF) (1998) *Microorganisms in Foods. 6. Microbial Ecology of Food Commodities*. Blackie Academic and Professional, London UK.

Ito, K. A. and Seeger, M. L. (1980) Effects of germicides on microorganisms in can cooling waters. *Journal of Food Protection*, **43**(6), 484-487.

Juneja, V. K., Snyder, O. P. and Cygnarowicz-Provost, M. (1994) Influence of cooling rate on outgrowth of *Clostridium perfringens* spores in cooked ground beef. *Journal of Food Protection*, **57**(12), 1063-1067.

Juneja, V. K., Snyder, O. P. and Marmer, B. S. (1997) Potential for growth from spores of *Bacillus cereus* and *Clostridium botulinum* and vegetative cells of *Staphylococcus aureus, Listeria monocytogenes*, and *Salmonella* serotypes in cooked ground beef during cooling. *Journal of Food Protection*, **60**(3), 272-275.

Karahadian, C., Lindsay, R. C., Dillman, L. L. *et al.* (1985) Evaluation of the potential for botulinal toxigenesis in reduced-sodium processed American cheese foods and spreads. *Journal of Food Protection*, **48**(1), 63-69.

Kautter, D. A., Lilly Jr., T., Lynt, R. K. *et al.* (1979) Toxin production by *Clostridium botulinum* in shelf-stable pasteurized process cheese spreads. *Journal of Food Protection*, **42**(10), 784-786.

Kautter, D. A., Lynt, R. K., Lilly Jr., T. *et al.* (1981) Evaluation of the botulism hazard from imitation cheeses. *Journal of Food Science*, **46**, 749-750, 764.

Kautter, D. A., Solomon, H. M. and Rhodehamel, E. J. (1992) *Clostridium botulinum*, in *Food and Drug Administration: Bacteriological Analytical Manual*, 7th edn. AOAC International, Arlington, Virginia, USA (Chapter 17), pp. 215-225.

Kim, J. and Foegeding, P. M. (1993) Principles of control, in Clostridium botulinum - *Ecology and Control in Foods* (eds A. H. W. Hauschild and K. L. Dodds). Marcel Dekker, Inc, New York, USA, pp. 121-176.

Korkeala, H., Stengel, G., Hyytiä, E. *et al.* (1998) Type E botulism associated with vacuum-packaged hot-smoked whitefish. *International Journal of Food Microbiology*, **43**, 1-5.

Kotev, S., Leventhal, A., Bashary, A. *et al.* (1987) International outbreak of type E botulism associated with ungutted, salted whitefish. *Morbidity and Mortality Weekly Report*, **36**(49), 812-813.

Kueper, T. V. and Trelease, R. D. (1974) Variables affecting botulinum toxin development and nitrosamine formation in fermented sausages. *Proceedings of the Meat Industry Conference*, American Meat Institute Foundation, Washington DC, USA, 69-74.

LaGrange Loving, A. (1998) Botulism in flavored oils - a review. *Dairy, Food and Environmental Sanitation*, **18**(6), 438-441.

Larson, A. E., Johnson, E. A., Barmore, C. R. *et al.* (1997) Evaluation of the botulism hazard from vegetables in modified atmosphere packaging. *Journal of Food Protection*, **60**(10), 1208-1214.

Larson, A. E. and Johnson, E. A. (1999) Research note - evaluation of botulinal toxin production in packaged fresh-cut Cantaloupe and Honeydew melons. *Journal of Food Protection*, **62**(8), 948-952.

Leighton, G. (1923) *Botulism and Food Preservation (The Loch Maree Tragedy)*. W. Collins Sons & Co Ltd, Glasgow, Scotland, UK.

Lilly, T. and Kautter, D. A. (1990) Outgrowth of naturally occurring *Clostridium botulinum* in vacuum-packaged fresh fish. *Journal of the Association of Official Analytical Chemists*, **73**(2), 211-212.

Lilly Jr, T., Solomon, H. M. and Rhodehamel, E. J. (1996) Incidence of *Clostridium botulinum* in vegetables packaged under vacuum or modified atmosphere. *Journal of Food Protection*, **59**(1), 59-61.

Lücke, F-K., Hechelmann, H. and Leistner, L. (1981) The relevance to meat products of psychrotrophic strains of *Clostridium botulinum*, in *Psychrotrophic Microorganisms in Spoilage and Pathogenicity* (eds T. A. Roberts, G. Hobbs,

J. H. B. Christian and N. Skovgaard). Academic Press, London, UK, pp. 491–497.

Lücke, F-K., Hechelmann, H. and Leistner, L. (1983) Fate of *Clostridium botulinum* in fermented sausages processed with or without nitrite. *Proceedings of the 29th European Meeting of Meat Research Workers*, Parma, Italy, pp. 403–409.

Lücke, F-K. and Roberts, T. A. (1993) Control in meat and meat products, in Clostridium botulinum - *Ecology and Control in Foods* (eds A. H. W. Hauschild and K. L. Dodds). Marcel Dekker, Inc, New York, USA, pp. 177–207.

McCroskey, L. M., Hatheway, C. L., Fenicia, L. *et al.* (1986) Characterization of an organism that produces type E botulinal toxin but which resembles *Clostridium butyricum* from the feces of an infant with type E botulism. *Journal of Clinical Microbiology*, **23**(1), 201–202.

McCroskey, L. M., Hatheway, C. L., Woodruff, B. A. *et al.* (1991) Type F botulism due to neurotoxigenic *Clostridium baratii* from an unknown source in an adult. *Journal of Clinical Microbiology*, **29**(11), 2618–2620.

MacDonald, K. L., Spengler, R. F., Hatheway, C. L. *et al.* (1985) Type A botulism from sautéed onions. *Journal of the American Medical Association*, **253**(9), 1275–1278.

MacDonald, K. L., Cohen, M. L. and Blake, P. A. (1986) The changing epidemiology of adult botulism in the United States. *American Journal of Epidemiology*, **124**, 794–799.

MAFF (1996) *Report on the National Study of Canneries*. (Number 5 in a joint series.) Ministry of Agriculture, Fisheries and Food and Department of Health, London, UK.

Mann, J. M., Lathrop, G. D. and Bannerman, J. A. (1983) Economic impact of a botulism outbreak. *Journal of the American Medical Association*, **249**(10), 1299–1301.

Meng, J. and Genigeorgis, C. A. (1993) Modelling lag phase of non-proteolytic *Clostridium botulinum* toxigenesis in cooked turkey and chicken breast as affected by temperature, sodium lactate, sodium chloride and spore inoculum. *International Journal of Food Microbiology*, **19**, 109–122.

Meng, J. and Genigeorgis, C. A. (1994) Delaying toxigenesis of *Clostridium botulinum* by sodium lactate in 'sous-vide' products. *Letters in Applied Microbiology*, **19**, 20–23.

Meyer, K. F. and Dubovsky, B. J. (1922a) The occurrence of the spores of *B. botulinus* in Belgium, Denmark, England, The Netherlands and Switzerland. VI. *Journal of Infectious Diseases*, **31**, 600–609.

Meyer, K. F. and Dubovsky, B. J. (1922b) The distribution of spores of *B. botulinus* in California. II. *Journal of Infectious Diseases*, **31**, 541–555.

Meyer, K. F. and Gunnison, J. B. (1929) Botulism due to home canned Bartlett pears. *Journal of Infectious Diseases*, **31**, 135–147.

Mitchell, C. A. (1900) *Flesh Foods, with Methods for their Chemical, Microscopical and Bacteriological Examination*. Charles Griffin and Co., London, UK.

Montville, T. J. (1982) Metabiotic effect of *Bacillus licheniformis* on *Clostridium botulinum*: implications for home-canned tomatoes. *Applied and Environmental Microbiology*, **44**(2), 334–338.

Morse, D. L., Pickard, L. K., Guzewich, J. J. *et al.* (1990) Garlic-in-oil associated botulism: episode leads to product modification. *American Journal of Public Health*, **80**(11) 1372–1373.

Morton, R. D., Scott, V. N., Bernard, D. T. *et al.* (1990) Effect of heat and pH on toxigenic *Clostridium butyricum*. *Journal of Food Science*, **55**(6), 1725-1727, 1739.

Mossel, D. A. A., Corry, J. E. L., Struijk, C. B. *et al.* (1995). *Essentials of the Microbiology of Foods. A Textbook for Advanced Studies*. John Wiley and Sons Ltd, Chichester, UK.

Nakano, H., Yoshikuni, Y., Hashimoto, H. *et al.* (1992) Detection of *Clostridium botulinum* in natural sweetening. *International Journal of Food Microbiology*, **16**, 117-121.

National Food Processors Association (1968) *Laboratory Manual for Food Canners and Processors. Volume 1, Microbiology and Processing*. The AVI Publishing Company, Inc, Westport, Connecticut, USA.

National Food Processors Association (1989) *Guidelines for the Development, Production, Distribution and Handling of Refrigerated Foods*. National Food Processors Association, Microbiology and Food Safety Committee, Washington, DC, USA.

National Food Processors Association (1995) *Canned Foods - Principles of Thermal Process Control, Acidification and Container Closure Evaluation*. National Food Processors Association, Washington, DC, USA.

Nevin, M. (1921) Botulism from cheese. *Journal of Infectious Diseases*, **28**(3), 226-231.

Nordal, J. and Gudding, R. (1975) The inhibition of *Clostridium botulinum* type B and E in salami sausage. *Acta Veternaria Scandinavica*, **16**(4), 537-548.

Notermans, S. H. W. (1993) Control in fruits and vegetables, in Clostridium botulinum - *Ecology and Control in Foods* (eds. A. H. W. Hauschild and K. L. Dodds). Marcel Dekker, Inc, New York, USA, pp. 233-260.

Notermans, S., Dufrenne, J. and Gerrits, J. P. G. (1989) Natural occurrence of *Clostridium botulinum* on fresh mushrooms (*Agaricus bisporus*). *Journal of Food Protection*, **52**(10), 733-736.

Odlaug, T. E. and Pflug, I. J. (1977) Thermal destruction of *Clostridium botulinum* spores suspended in tomato juice in aluminium thermal death time tubes. *Applied and Environmental Microbiology*, **34**(1), 23-29.

Odlaug, T. E. and Pflug, I. J. (1977) Effect of storage time and temperature on the survival of *Clostridium botulinum* spores in acid media. *Applied and Environmental Microbiology*, **34**(1), 30-33.

Odlaug, T. E. and Pflug, I. J. (1978) *Clostridium botulinum* and acid foods. *Journal of Food Protection*, **41**(7), 566-573.

Odlaug, T. E. and Pflug, I. J. (1979) *Clostridium botulinum* growth and toxin production in tomato juice containing *Aspergillus gracilis*. *Applied and Environmental Microbiology*, **37**(3), 496-504.

O'Mahony, M., Mitchell, E., Gilbert, R. J. *et al.* (1990) An outbreak of foodborne botulism associated with contaminated hazelnut yoghurt. *Epidemiology and Infection*, **104**, 389-395.

Peck, M. W. (1999) Safety of sous-vide foods with respect to *Clostridium botulinum*. *Proceedings of the Third European Symposium on Sous-Vide*, 25-26 March, 1999. Katholieke Universiteit Leuven, Belgium.

Peterson, M. E., Pelroy, G. A., Poysky, F. T. *et al.* (1997) Heat-pasteurization process for inactivation of non-proteolytic types of *Clostridium botulinum* in picked Dungeness crabmeat. *Journal of Food Protection*, **60**(8), 928-934.

Petran, R. L., Sperber, W. H. and Davis, A. B. (1995) *Clostridium botulinum* toxin formation in Romaine lettuce and shredded cabbage: effect of storage and packaging conditions. *Journal of Food Protection*, **58**(6) 624-627.

Pivnick, H. and Bird, H. (1965) Toxinogenesis by *Clostridium botulinum* types A and E in perishable cooked meats vacuum-packed in plastic pouches. *Food Technology*, July, 132–140.

Potter, M. D., Meng, J. and Kimsey, P. (1993) An ELISA for detection of botulinal toxin types A, B, and E in inoculated food samples. *Journal of Food Protection*, **56**(10), 856–861.

Put, H. M. C., Witvoet, H. J. and Warner, W. R. (1980) Mechanism of microbiological leaker spoilage of canned foods: biophysical aspects. *Journal of Food Protection*, **43**(6), 488–497.

Raatjes, G. J. M. and Smelt, J. P. P. M. (1979) *Clostridium botulinum* can grow and form toxin at pH values lower than 4.6. *Nature*, **281** (4 Oct), 398–399.

Read Jr., R. B., Bradshaw, J. G. and Francis, D. W. (1970) Growth and toxin production of *Clostridium botulinum* type E in milk. *Journal of Dairy Science*, **53**(9), 1183–1186.

Reddy, N. R., Solomon, H. M., Yep, H. *et al.* (1997) Shelf life and toxin development by *Clostridium botulinum* during storage of modified-atmosphere-packaged fresh aquacultured salmon fillets. *Journal of Food Protection*, **60**(9), 1055–1063.

Rhodehamel, E. J., Reddy, N. R. and Pierson, M. D. (1992) Botulism: the causative agent and its control in foods. *Food Control*, **3**(3), 125–142.

Roberts, T. A. and Ingram, M. (1973) Inhibition of growth of *C. botulinum* at different pH values by sodium chloride and sodium nitrite. *Journal of Food Technology*, **8**, 467–475.

Robins, M., Brocklehurst, T. and Wilson, P. (1994) Food structure and the growth of pathogenic bacteria. *Food Technology International Europe*, 31–36.

Ryan, M., Webb Jr., C. R., Mann, J. M. *et al.* (1978) Botulism – New Mexico. *Morbidity and Mortality Weekly Report*, **27**, 138.

St Louis, M. E., Peck, S. H. S., Bowering, D. *et al.* (1988) Botulism from chopped garlic: delayed recognition of a major outbreak. *Annals of Internal Medicine*, **108**(3), 363–368.

Salyers, A. A. and Whitt, D. D. (1994) *Bacterial Pathogenesis – A Molecular Approach*. ASM Press, Washington, DC, USA, pp. 130–136.

Schiavo, G., Rossetto, O., Santucci, A. *et al.* (1992a) Botulinum neurotoxins are zinc proteins. *The Journal of Biological Chemistry*, **267**(33), 23479–23483.

Schiavo, G., Benfenati, F., Poulain, B. *et al.* (1992b) Tetanus and botulinum-B neurotoxins block neurotransmitter release by proteolytic cleavage of synaptobrevin. *Nature*, **359**, 832–835.

Schiavo, G., Malizio, C., Trimble, W. S. *et al.* (1994) Botulinum G neurotoxin cleaves VAMP/Synoptobrevin at a single Ala–Ala peptide bond. *The Journal of Biological Chemistry*, **269**(32), 20213–20216.

Schmidt, C. F., Lechowich, R. V. and Folinazzo, J. F. (1961) Growth and toxin production by type E *Clostridium botulinum* below 40°F. *Journal of Food Science*, **26**, 626–630.

Seals, J. E., Snyder, J. D., Edell, T. A. *et al.* (1981) Restaurant-associated type A botulism: transmission by potato salad. *American Journal of Epidemiology*, **113**(4), 436–444.

Shone, C. C. (1987) Understanding toxin action: *Clostridium botulinum* neurotoxins, their structures and modes of action, in *Natural Toxicants in Foods* (ed. D.H. Watson). Ellis Horwood Ltd, Chichester, UK, pp. 11–57.

Shone, C. C., Wilton-Smith, P., Appleton, N. *et al.* (1985) Monoclonal antibody-based immunoassay for type A *Clostridium botulinum* toxin is comparable to the mouse bioassay. *Applied and Environmental Microbiology*, **50**(1), 63–67.

Silas, J. C., Carpenter, J. A., Hamdy, M. K. *et al.* (1985) Selective and differential medium for detecting *Clostridium botulinum. Applied and Environmental Microbiology*, **50**, 1110-1111.

Simini, B. (1996) Outbreak of foodborne botulism continues in Italy. *The Lancet*, **348** (21 Sept), 813.

Simpson, M. V., Smith, J. P., Dodds, K. *et al.* (1995) Challenge studies with *Clostridium botulinum* in a sous-vide spaghetti and meat-sauce product. *Journal of Food Protection*, **58**(3), 229-234.

Singleton, P. and Sainsbury, D. (1987) *Dictionary of Microbiology and Molecular Biology*, 2nd edn. John Wiley & Sons Ltd, Chichester, UK.

Smelt, J. P. P. M., Raatjes, G. J. M., Crowther, J. S. *et al.* (1982) Growth and toxin formation by *Clostridium botulinum* at low pH values. *Journal of Applied Bacteriology*, **52**, 75-82.

Smith, G. R. and Young, A. M. (1980) *Clostridium botulinum* in British soil. *Journal of Hygiene, Cambridge*, **85**, 271-274.

Smith, G. R., Milligan, R. A. and Moryson, C. J. (1978) *Clostridium botulinum* in aquatic environments in Great Britain and Ireland. *Journal of Hygiene, Cambridge*, **80**, 431-438.

Smith, L. DS. (1977) *Botulism, the Organism, its Toxins, the Disease.* Charles C. Thomas, Illinois, USA.

Snyder, O. P. (1996) Redox potential in deli foods: botulism risk? *Dairy, Food and Environmental Sanitation*, **16**(9), 546-548.

Solomon, H. M. and Kautter, D. A. (1986) Growth and toxin production by *Clostridium botulinum* in sautéed onions. *Journal of Food Protection*, **49**(8), 618-620.

Solomon, H. M. and Kautter, D. A. (1988) Outgrowth and toxin production by *Clostridium botulinum* in bottled chopped garlic. *Journal of Food Protection*, **51**(11), 862-865.

Solomon, H. M., Kautter, D. A., Rhodehamel, E. J. *et al.* (1991) Evaluation of un-acidified products bottled in oil for outgrowth and toxin production by *Clostridium botulinum. Journal of Food Protection*, **54**(8), 648-649.

Solomon, H. M., Rhodehamel, E. J. and Kautter, D. A. (1994) Growth and toxin production by *Clostridium botulinum* in sliced raw potatoes under vacuum with and without sulfite. *Journal of Food Protection*, **57**(10), 878-881.

Solomon, H. M., Rhodehamel, E. J. and Kautter, D. A. (1998) Growth and toxin production by *Clostridium botulinum* on sliced raw potatoes in a modified atmosphere with and without sulfite. *Journal of Food Protection*, **61**(1), 126-128.

Somers, E. B. and Taylor, S. L. (1987) Antibotulinal effectiveness of nisin in pasteurized process cheese spreads. *Journal of Food Protection*, **50**(10), 842-848.

Stersky, A., Todd, E. and Pivnick, H. (1980) Food poisoning associated with post-process leakage (PPL) in canned foods. *Journal of Food Protection*, **43**(6), 465-476.

Stringer, S. C. and Peck, M. W. (1997) Combinations of heat treatment and sodium chloride that prevent growth from spores of non-proteolytic *Clostridium botulinum. Journal of Food Protection*, **60**(12), 1553-1559.

Sugiyama, H. (1980) *Clostridium botulinum* neurotoxin. *Microbiological Reviews*, **44**(3), 419-448.

Sugiyama, H. (1982) Botulism hazards from non-processed foods. *Food Technology* (Dec.), 113-115.

Sugiyama, H. and Rutledge, K. S. (1978) Failure of *Clostridium botulinum* to grow in fresh mushrooms packaged in plastic film overwraps with holes. *Journal of Food Protection*, **41**(5), 348-350.

Sugiyama, H., Woodburn, M., Yang, K. H. *et al.* (1981) Production of botulinum toxin in inoculated pack studies of foil-wrapped baked potatoes. *Journal of Food Protection*, **44**(12), 896-898.

Szabo, E. A., Pemberton, J. M. and Desmarchelier, P. M. (1993) Detection of the genes encoding botulinum neurotoxin types A to E by the polymerase chain reaction. *Applied and Environmental Microbiology*, **59**(9), 3011-3020.

Szabo, E. A., Pemberton, J. M., Gibson, A. M. *et al.* (1994) Polymerase chain reaction for detection of *Clostridium botulinum* types A, B and E in food, soil and infant faeces. *Journal of Applied Bacteriology*, **76**, 539-545.

Tanaka, N. (1982) Challenge of pasteurized process cheese spreads with *Clostridium botulinum* using in-process and post-process inoculation. *Journal of Food Protection*, **45**(11), 1044-1050.

Tanaka, N., Traisman, E., Plantinga, P. *et al.* (1986) Evaluation of factors involved in anti-botulinal properties of pasteurized process cheese spreads. *Journal of Food Protection*, **49**(7), 526-531.

Telzak, E. E., Bell, E. P., Kautter, D. A. *et al.*(1990) An international outbreak of type E botulism due to un-eviscerated fish. *Journal of Infectious Diseases*, **161**, 340-342.

Terranova, W., Breman, J. G., Locey, R. P. *et al.* (1978) Botulism type B: epidemiologic aspects of an extensive outbreak. *American Journal of Epidemiology*, **108**(2), 150-156.

Ter Steeg, P. F. and Cuppers, H. G. A. M. (1995) Growth of proteolytic *Clostridium botulinum* in process cheese products: II. Predictive modelling. *Journal of Food Protection*, **58**(10), 1100-1108.

Ter Steeg, P. F., Cuppers, H. G. A. M., Hellemons, J. C. *et al.* (1995) Growth of proteolytic *Clostridium botulinum* in process cheese products: I. Data acquisition for modelling the influence of pH, sodium chloride, emulsifying salts, fat dry basis, and temperature. *Journal of Food Protection*, **58**(10), 1091-1099.

Thatcher, F. S., Erdman, I. E. and Pontefract, R. D. (1967) Some laboratory and regulatory aspects of the control of *C. botulinum* in processed foods, in *Botulism 1966* (eds M. Ingram and T. A. Roberts). Chapman and Hall, London, UK, pp. 511-521.

Tompkin, R. B. (1980) Botulism from meat and poultry products – a historical perspective. *Food Technology* (May), 229-236, 257.

Tompkin, R. B., Christiansen, L. N. and Shaparis, A. B. (1978) Enhancing nitrite inhibition of *Clostridium botulinum* with isoascorbate in perishable canned cured meat. *Applied and Environmental Microbiology*, **35**(1), 59-61.

Tompkin, R. B., Christiansen, L. N. and Shaparis, A. B. (1979a) Isoascorbate level and botulinal inhibition in perishable canned cured meat. *Journal of Food Science*, **44**, 1147-1149.

Tompkin, R. B., Christiansen, L. N. and Shaparis, A. B. (1979b) Iron and the anti-botulinal efficacy of nitrite. *Applied and Environmental Microbiology*, **37**(2), 351-353.

Topley, W. W. C. and Wilson, G. S. (1929) *The Principles of Bacteriology and Immunity*, Volume II. Edward Arnold and Co, London, UK.

Townes, J. M., Cieslak, P. R., Hatheway, C. L. *et al.* (1996) An outbreak of type A botulism associated with a commercial cheese sauce. *Annals of Internal Medicine*, **125**(7), 558-563.

Tuynenburg Muys, G. (1971) Microbial safety in emulsions. *Process Biochemistry*, 6(6), 25-28.

United States Department of Agriculture (1998) Requirements for the production of cooked beef, roast beef, and cooked corned beef. *Code of Federal Regulations, Title 9 - Animals and Animal Products, Section 318.17*. Food Safety and Inspection Service, Washington, DC, USA.

Wictome, M., Newton, K., Jameson, K. *et al.* (1999) Development of an *in vitro* bioassay for *Clostridium botulinum* type B neurotoxin in foods that is more sensitive than the mouse bioaasay. *Applied and Environmental Microbiology*, 65(9), 3787-3792.

Yamasaki, S., Baumeister, A., Binz, T. *et al.* (1994) Cleavage of members of the synaptobrevin/VAMP family by types D and F botulinal neurotoxins and tetanus toxin. *Journal of Biological Chemistry*, 269(17), 12764-12772.

INDEX